Environment: Problems and Solutions

G. Tyler Miller, Jr.
Adjunct Professor of Human Ecology
St. Andrews Presbyterian College

Wadsworth Publishing Company
Belmont, California
A Division of Wadsworth, Inc.

Biology and Environmental Science Publisher: Jack Carey
Editorial Assistant: Kristin Milotich
Development Editors: Mary Arbogast and Autumn Stanley
Production Editor: Carol Carreon Lombardi
Managing Designer: Andrew H. Ogus
Print Buyer: Diana Spence
Art Editor: Donna Kalal
Permissions Editor: Peggy Meehan
Copy Editor: Alan Titche
Technical Illustrators: Darwin and Vally Hennings; Tasa Graphic Arts, Inc.; Susan Breitbard; Teresa Roberts; Carlyn Iverson; Cindie Wooley
Box Logo Designs: The Weller Institute
Electronic Composition: Brandon Carson, Wadsworth Digital Productions
Cover Design: Andrew H. Ogus
Printer: Arcata Book Group/Kingsport
Cover Photograph: Endangered Florida panther, Art Wolfe/AllStock

Thanks to the Center for Plant Conservation for providing source material for the illustration of the Knowlton cactus on p. 71.

This book is printed on acid-free recycled paper.

International Thomson Publishing
The trademark ITP is used under license.

Printed in the United States of America
1 2 3 4 5 6 7 8 9 10—98 97 96 95 94

Library of Congress Cataloging-in-Publication Data
Miller, G. Tyler (George Tyler), 1931–
Environment: problems and solutions / G. Tyler Miller, Jr.
p. cm.
Includes index.
ISBN 0-534-23394-5
1. Environmental sciences. 2. Environmental policy. I. Title.
GE105.M54 1994 93-47669
363.20—dc20

Books in the Wadsworth Biology Series

Environment: Problems and Solutions, Miller
Environmental Science, 4th, Miller
Living in the Environment, 8th, Miller
Resource Conservation and Management, Miller
Sustaining the Earth, Miller
Biology: Concepts and Applications, 2nd, Starr
Biology: The Unity and Diversity of Life, 6th, Starr and Taggart
Dimensions of Cancer, Kupchella
Evolution: Process and Product, 3rd, Dodson and Dodson
Biology of the Cell, 2nd, Wolfe
Biology: The Foundations, 2nd, Wolfe
Introduction to Cell Biology, Wolfe
Molecular and Cellular Biology, Wolfe
Oceanography: An Introduction, 4th, Ingmanson and Wallace
Oceanography: An Invitation to Marine Science, Garrison
Plant Physiology, 4th, Devlin and Witham
Exercises in Plant Physiology, Witham et al.
Plant Physiology, 4th, Salisbury and Ross
Plant Physiology Laboratory Manual, Ross
Plants: An Evolutionary Survey, 2nd, Scagel et al.
Psychobiology: The Neuron and Behavior, Hoyenga and Hoyenga
Sex, Evolution, and Behavior, 2nd, Daly and Wilson

Of Related Interest:

Environmental Ethics, DesJardins
The Environmental Ethics and Policy Book, VanDeVeer and Pierce
Radical Environmentalism, List
Watersheds: Classic Cases in Environmental Ethics, Newton and Dillingham

Two trees have been planted in a tropical rain forest for every tree used to make this book, courtesy of G. Tyler Miller, Jr., and Wadsworth Publishing Company. The author also sees that 50 trees are planted to compensate for the paper he uses and that several hectares of tropical rain forest are protected.

PREFACE

For Instructors and Students

This brief introduction to environmental problems and solutions was written as a primer on environmental science for a wide range of courses in the college curriculum. After Chapter 1 has been covered, the rest of the book can be used in almost any order. In addition, many sections within chapters can be moved around or omitted to accommodate courses with different lengths and emphases.

This book uses basic scientific laws, principles, and concepts to help us understand environmental and resource problems (see *Summary of Principles* on pp. 16–17) and the possible solutions to these problems.

I also relate the information in the book to the real world and to our individual lives, in both the main text and the boxes sprinkled throughout the book. These include: **(1)** *Spotlights* highlighting and giving further insights into environmental problems and concepts; **(2)** *Case Studies* giving in-depth information about key issues; **(3)** *Connections* showing how various environmental problems, concepts, and solutions are interrelated; **(4)** *Solutions* summarizing possible solutions to environmental problems or describing what individuals have done to help sustain the earth for us and all life; and **(5)** *Individuals Matter* giving examples of what we as individuals can do to help sustain the earth.

The book's 90 illustrations are designed to present complex ideas in understandable ways and to relate learning to the real world. Instead of using photos I have converted important ideas to drawings.

Factual recall questions (with answers) are listed at the bottom of most pages. Each chapter ends with a set of questions designed to encourage critical thinking.

To reduce student costs this book is printed in black and white and has a soft cover. It is also printed on acid-free recycled paper with the highest available post-consumer paper waste.

Tell me how you think this book can be improved, and if you find any errors, please let me know about them. Send any errors you find and your suggestions for improvement to Jack Carey, Science Editor, Wadsworth Publishing Company, 10 Davis Drive, Belmont, CA 94002. He will send them on to me.

The following supplements are available:

- *Green Lives, Green Campuses: An Activities Workbook* written by Jane Heinze-Fry. This workbook is designed to help students apply environmental concepts by investigating their lifestyles and by making an environmental audit of their campus.
- *Instructor's Manual,* by Jane Heinze-Fry, contains teaching suggestions, activities, class projects, and an extensive test bank.
- A set of 387 black-and-white transparency masters for making overhead transparencies is available to adopters.
- A special version of STELLA II software, a tool for developing critical thinking, is available, together with an accompanying workbook.

I wish to thank the members of Wadsworth's talented production team, listed on the copyright page, who have made important contributions. My thanks also go to Jane Heinze-Fry for her outstanding work on *Green Lives, Green Campuses* and on the *Instructor's Manual.*

Special thanks go to Jack Carey, Science Editor at Wadsworth, for his encouragement, help, friendship, and superb reviewing system. It helps immensely to work with the best and most experienced editor in college textbook publishing.

I also wish to thank Peggy Sue O'Neal, my spouse and best friend, for her love and support of me and the earth. I dedicate this book to her and to the earth that sustains us all.

G. Tyler Miller, Jr.

Contents

1 Environmental Problems and Their Causes

We need to come together and choose a new direction. We need to transform our society into one in which people live in true harmony—harmony among nations, harmony among the races of humankind, and harmony with nature.... We will either reduce, reuse, recycle, and restore—or we will perish.

REV. JESSE JACKSON

Living in an Exponential Age

EXPONENTIAL GROWTH Once there were two kings who enjoyed playing chess, with the winner claiming a prize from the loser. After one match the winning king asked the loser to place one grain of wheat on the first square of the chessboard, two on the second, four on the third, and so on. The number of grains was to double each time until all 64 squares were filled.

The losing king, thinking he was getting off easy, agreed with delight. It was the biggest mistake he ever made. He bankrupted his kingdom and still could not produce the 2^{64} grains of wheat he had promised. In fact, it's probably more than all the wheat that has ever been harvested! This is an example of **exponential growth**, in which a quantity increases by a fixed percentage of the whole in a given time. As the losing king learned, exponential growth is deceptive: It starts off slowly, but after only a few doublings it rises to enormous numbers because each doubling is more than the total of all earlier growth.

Here is another example. Fold a piece of paper in half to double its thickness. If you could manage to do this 42 times, the stack would reach from the earth to the moon, 386,400 kilometers (240,000 miles) away. If you could double it 50 times, the folded paper would almost reach the sun, 149 million kilometers (93 million miles) away!

The environmental problems we face—population growth, excessive and wasteful use of resources, extinction of plants and animals, and pollution—are interconnected and growing exponentially. For example, world population has more than doubled in only 43 years, from 2.5 billion in 1950 to 5.5 billion in 1993. Unless death rates rise sharply or birth rates drop sharply, it may reach 11 billion by 2045 and 14 billion by 2100 (Figure 1-1).

Exponential growth in population and in resource use (Figure 1-2) has drastically changed the face of the planet. Each year more forests, grasslands, and wetlands disappear, and some deserts grow larger. Vital topsoil, washed or blown away from farmland and cleared forests, clogs streams, lakes, and reservoirs with sediment. Underground water is pumped from wells faster than it can be replenished. Oceans are used as trash cans for many of our wastes. Every hour we drive as many as four wildlife species to extinction.

Burning fossil fuels and cutting down and burning forests raise the concentrations of carbon dioxide and other heat-trapping gases in the lower atmosphere. Within the next 40 to 50 years the earth's climate may become warm enough to disrupt agricultural productivity, alter water distribution, and drive countless species to extinction.

Extracting and burning fossil fuels pollutes the air and water and disrupts the land. Other chemicals we add to the air drift into the upper atmosphere and deplete ozone gas, which filters out much of the sun's harmful ultraviolet radiation. Toxic wastes from factories and homes poison the air, water, and soil. Agricultural pesticides contaminate some of our drinking water and food.

EARTH CAPITAL AND LIVING SUSTAINABLY Our existence, lifestyles, and economies depend totally on the sun and the earth. We can think of energy from the sun as **solar capital** and we can think of the planet's air, water, fertile soil, forests, grasslands, wetlands, oceans, streams, lakes, wildlife, minerals, and natural purification and recycling processes as **Earth capital**. Scientific research has revealed that the earth's life-support system for us and other species is made up of several layers or spheres (Figure 1-3):

- The **atmosphere**—a thin envelope of air around the planet. Its inner layer, the **troposphere**, extends only about 17 kilometers (11 miles) above sea level but contains most of the planet's air—

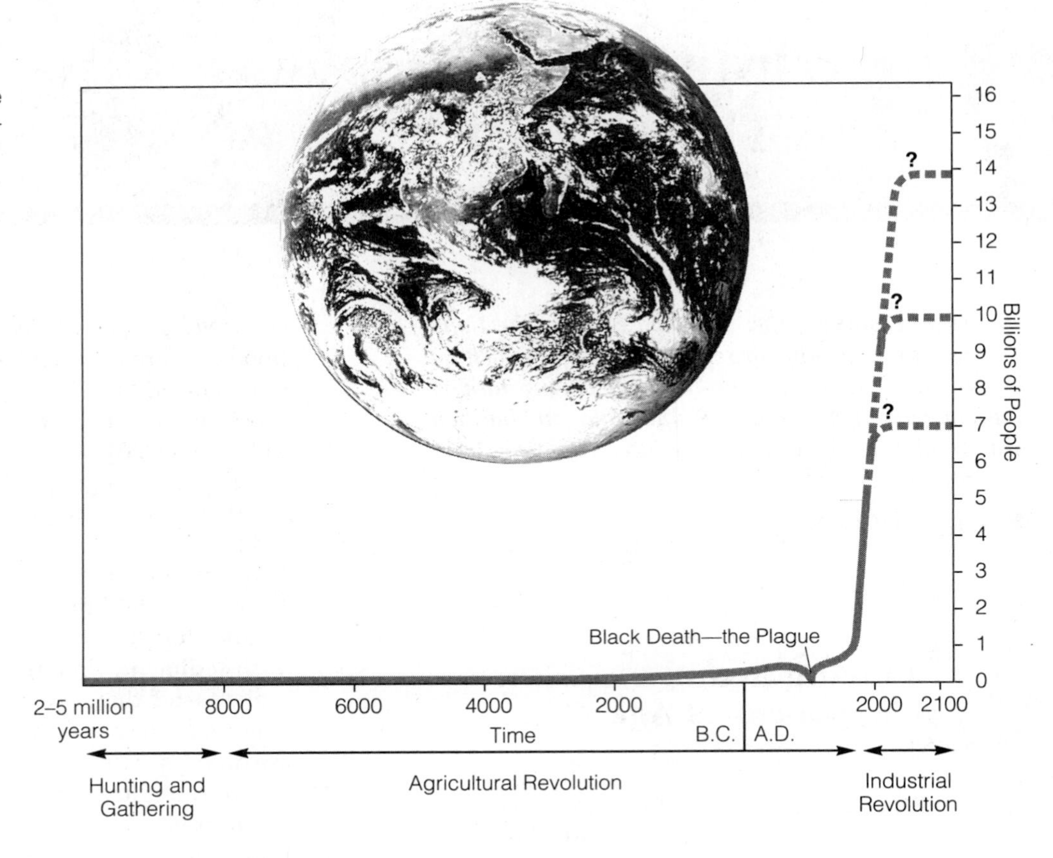

Figure 1-1 The J-shaped curve of past exponential world population growth, with projections to beyond 2100. Notice that exponential growth starts off slowly, but as time passes the curve becomes increasingly steeper. (Figure not to scale.) (Data from World Bank and United Nations)

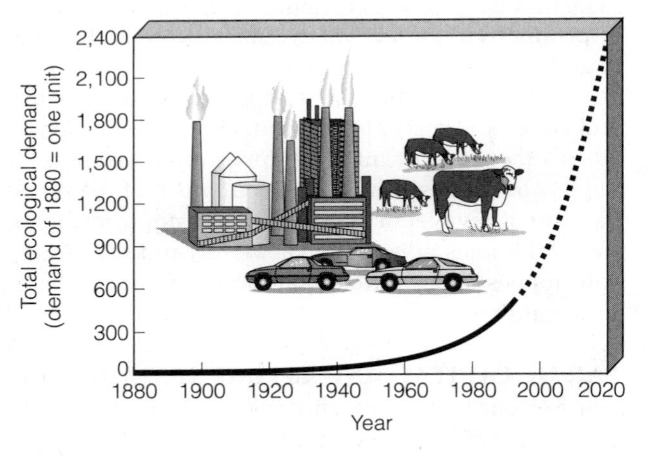

Figure 1-2 The J-shaped curve of exponential growth in the total ecological demand on the earth's resources from agriculture, mining, and industry between 1880 and 1993. Projections to 2020 assume that resource use will continue to grow at the current rate of 5.5% per year. At that rate our total ecological demand on the earth's resources doubles every 13 years. If global economic output grew by only 3% a year, resource consumption would still double every 23 years. (Data from United Nations, World Resources Institute, and Carrol Wilson, *Man's Impact on the Global Environment*, Cambridge, Mass.: MIT Press, 1970)

mostly nitrogen (78%) and oxygen (21%). The next layer, stretching 17–48 kilometers (11–30 miles) above the earth's surface, is called the **stratosphere**. Its lower portion contains enough ozone (O_3) to filter out most of the sun's harmful ultraviolet radiation, thus allowing life on land to exist.

- The **hydrosphere**—liquid water (both surface and underground), frozen water (polar ice, icebergs, permafrost in soil), and water vapor in the atmosphere.
- The **lithosphere**—the earth's crust and upper mantle. It contains the fossil fuels and minerals we use and the soil chemicals (nutrients) needed to support plant life.
- The **ecosphere** or **biosphere**—the portion of the earth where living (biotic) organisms exist and interact with one another and with their nonliving (abiotic) environment. This zone reaches from the deepest ocean floor 20 kilometers (12 miles) below sea level to the tops of the highest mountains. If the earth were an apple, the ecosphere—a haven for life between the earth's molten interior and the cold lifeless void of space—would be no thicker than the apple's skin.

Q: How many people are there in the world?

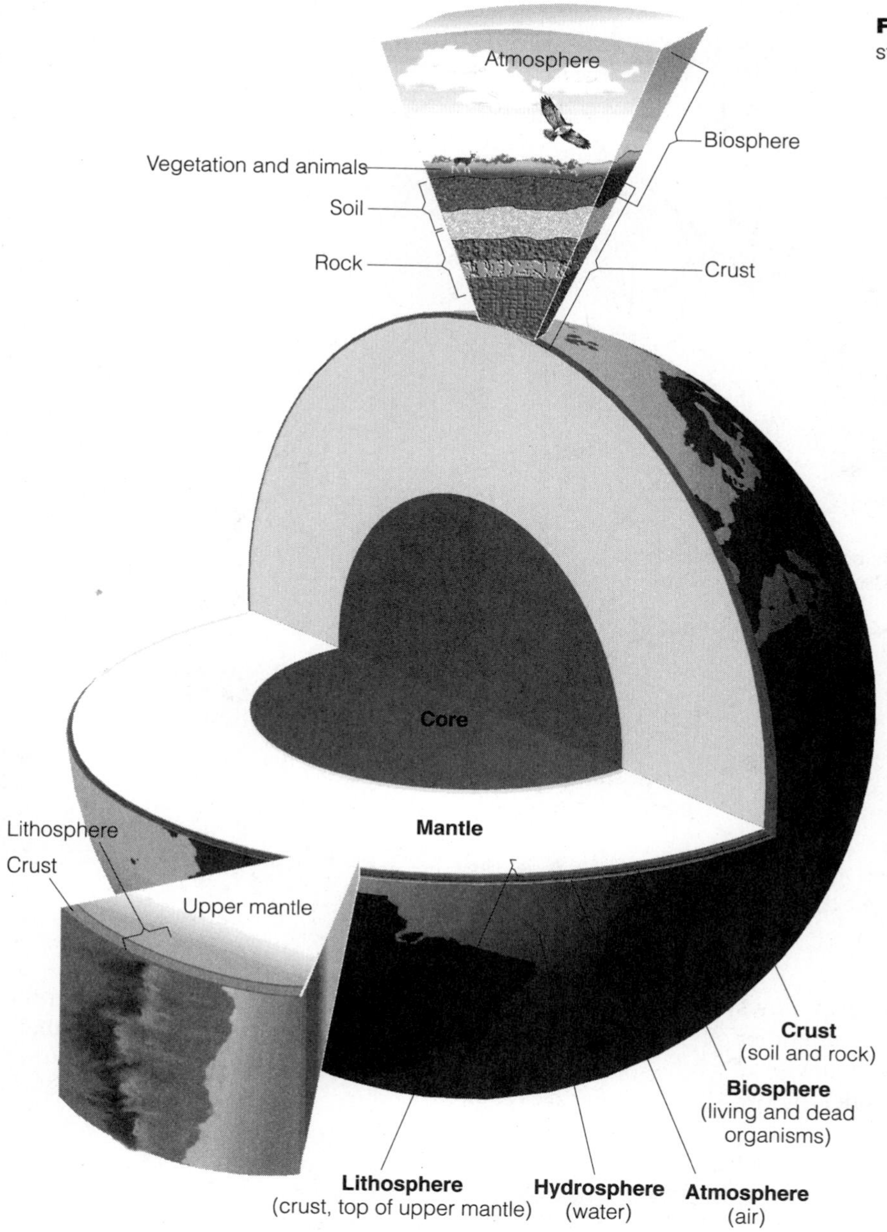

Figure 1-3 Our life-support system: the general structure of the earth.

Life on Earth depends on three connected factors (Figure 1-4):

- The *one-way flow of high-quality (usable) energy* from the sun, first through materials and living things in food chains and webs on or near the earth's surface, then into the environment as low-quality energy (mostly heat dispersed into air or water molecules at a low temperature), and eventually back into space.
- The *cycling of matter* required as *nutrients* by living organisms through parts of the ecosphere. The earth's chemical cycles connect past, present, and future forms of life. Thus, some of the carbon atoms in the skin of your right hand may once have been part of a leaf, a dinosaur's skin, or a layer of limestone rock. And some of the oxygen molecules you just inhaled may have been inhaled by your grandmother, by Plato, or by a hunter-gatherer who lived 25,000 years ago.
- *Gravity*, caused mostly by the attraction between the sun and the earth, which allows the planet to hold onto its atmosphere and causes the downward movement of chemicals in the matter cycles.

The basic problem we face is that we are depleting and degrading the earth's natural capital at an accelerating rate (Figures 1-1 and 1-2), and in the

A: Approximately 5.5 billion, as of mid-1993 (0.1 billion added per year)

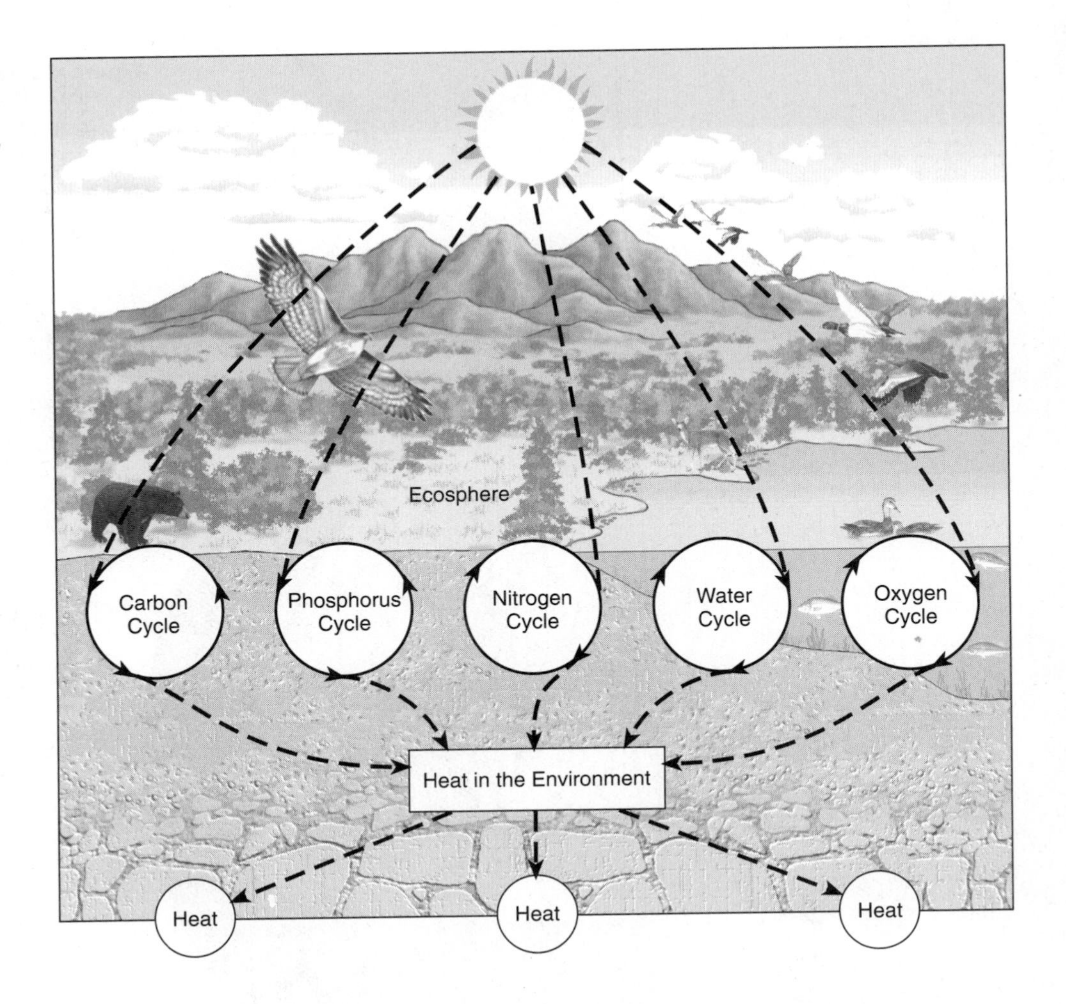

Figure 1-4 Life on the earth depends on the *one-way flow of energy* (dashed lines) from the sun through the ecosphere, the *cycling of critical elements* (solid lines around circles), and *gravity*, which keeps atmospheric gases from escaping into space and draws chemicals downward in the matter cycles. This simplified overview shows only a few of the many cycling elements.

process we are disrupting some of the energy flows and matter cycles (Figure 1-4) that keep us and other species alive. To environmentalists, such behavior is unsustainable. They argue that we need to learn how to live sustainably by working with the earth.

A **sustainable society** manages its economy and population size without doing irreparable environmental harm. It satisfies the needs of its people without depleting Earth capital and thereby jeopardizing the prospects of future generations of humans or other species. This is done by

- regulating population growth
- taking no more potentially renewable resources from the natural world than can be replenished naturally
- not overloading the capacity of the environment to cleanse and renew itself by natural processes
- encouraging Earth-sustaining rather than Earth-degrading forms of economic development
- reducing poverty (which can cause people to use land and other resources unsustainably for short-term survival)

The great news is that we can help sustain the earth for human beings and other species indefinitely by learning how to live off Earth income instead of Earth capital.

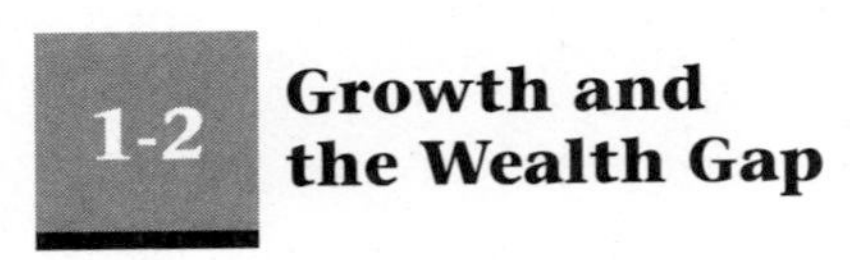

1-2 Growth and the Wealth Gap

HUMAN POPULATION GROWTH The increasing size of the human population is one example of exponential growth. As the population base grows, the number of people on the earth soars. The population growth curve rounds a bend and heads almost straight up, creating a J-shaped curve (Figure 1-1).

It took 2 million years to reach a billion people; 130 years to add the second billion; 30 years for the third; 15 years for the fourth; and only 12 years for the fifth billion. At current growth rates the sixth billion will be added during the 11-year period between 1987 and 1998, and the seventh only 10 years later, in 2008.

Q: How many people are added to the world's population each day?

Economics to the Rescue: Full-Cost Pricing

SOLUTIONS

Making, distributing, or using any economic good or service involves **external costs**—costs that are not included in the market price. Harmful external costs such as pollution and environmental degradation are passed on to workers, to the general public, and in some cases to future generations. Because these harmful costs aren't included in the market price, we don't usually connect them with the things we buy. As consumers and taxpayers, however, we pay these hidden costs sooner or later in the form of higher taxes, higher health costs, higher health insurance premiums, and higher cleaning and maintenance bills.

One way to deal with the problem of harmful external costs is for the government to add taxes, pass laws, provide subsidies, or use other strategies that force or entice producers to include all or most of the external costs in the market price of economic goods and services. Then the market price would be closer to the **full cost** of a good or service.

This approach requires government action because few companies will increase their cost of doing business by internalizing their external costs unless their competitors must do so as well.

What would happen if such a policy were phased in over the next 10 to 20 years? Economic growth would be redirected to reward the production of Earth-sustaining goods and services and to discourage Earth-degrading ones. We would pay more for most things because their market prices would be closer to their true costs, but everything would be "up front"; external costs would no longer be hidden.

Moreover, some things might even cost less. Internalizing external costs encourages producers to find ways both to cut costs (by inventing more resource-efficient and less environmentally harmful methods of production), and to offer less harmful, Earth-sustaining (or *green*) products. Jobs would be lost in Earth-degrading businesses, but at least as many jobs—probably more—could be created in Earth-sustaining businesses. Because businesses go where the profits are, many of today's Earth-degrading businesses would become the Earth-sustaining businesses of the future.

As external costs are internalized, however, governments must reduce income and other taxes and must withdraw subsidies formerly used to hide and pay for these external costs. Otherwise, consumers will face higher market prices without tax relief—a policy guaranteed to fail.

It's difficult to internalize external costs because it's not easy to put a price tag on every harmful effect of providing and using an economic good or service. People also disagree on the values they attach to various costs and benefits, and those in economic and political power find it difficult to change the rewards that have given them that power. Initiating full-cost pricing won't be easy, but such a change in the way we do business can occur if enough citizens become politically active and insist that it be done.

The relentless ticking of this population clock means that in 1993 the world's population of 5.5 billion people grew by about 90 million—an average increase of 247,000 people a day, or 10,300 people an hour.

ECONOMIC GROWTH Virtually all countries seek **economic growth**—an increase in their capacity to provide goods and services for people's final use. Such growth is accomplished by maximizing the flow of matter and energy resources through an economy (throughput)—by means of population growth (more consumers), more consumption per person, or both.

Most economists and investors argue that we must have unlimited economic growth to create jobs, satisfy people's economic needs and wants, clean up the environment, and help reduce poverty. They see the earth as an essentially unlimited source of raw materials and the environment as a virtually infinite sink for wastes. Limited resource supplies can be overcome by technological innovation.

On the other hand, environmentalists and some economists argue that economic systems depend on natural resources (ultimately provided by the sun and by the earth's natural processes), and that sooner or later depleting this Earth capital leads to unsustainable societies. If these contentions are correct, over the next few decades we must replace the economics of unlimited growth with the economics of sustainability. One method that can help us make this change is full-cost pricing of goods and services (Solutions, above).

RICH AND POOR COUNTRIES: THE WEALTH GAP The United Nations broadly classifies the world's countries as more developed or less developed according to their relative position on the ladder of industrialization and economic growth. The **more**

A: 247,000

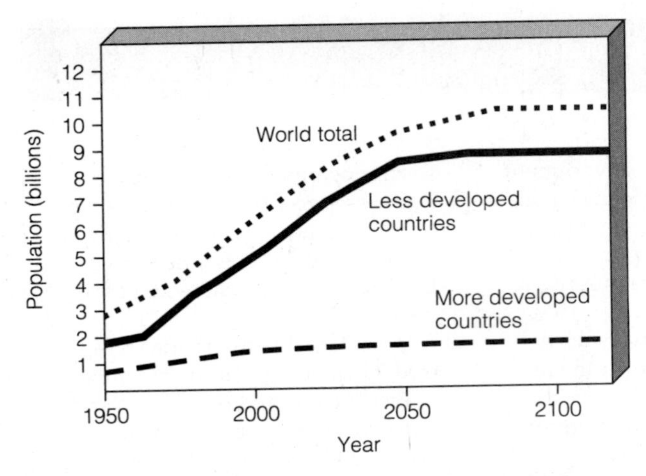

Figure 1-5 Past and projected population size for MDCs, LDCs, and the world, 1950–2120. (Data from United Nations)

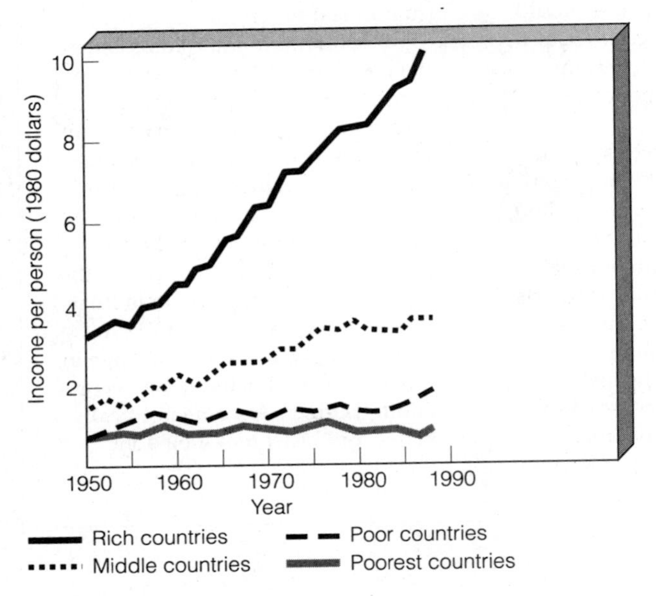

Figure 1-6 The wealth gap: changes in the distribution of global income per person in four types of countries, 1950–1990. Instead of trickling down, most of the income from economic growth has flowed up, with the situation worsening since 1980. (Data from United Nations)

developed countries (MDCs), or *developed countries*, include the United States, Canada, Japan, the former USSR, Australia, New Zealand, and the western-European countries. These countries, with 1.2 billion people (22% of the world's population), command about 85% of the world's wealth and income, use 88% of its natural resources and 73% of its energy, and generate most of its pollution and wastes.

All other nations are classified as **less developed countries (LDCs)**, or *developing countries*, with low to moderate industrialization and economic development. Most lie in Africa, Asia, and Latin America. Their 4.3 billion people, or 78% of the world's population, have only about 15% of the wealth and income, and they use only about 12% of the natural resources and 27% of the energy.

Most of the projected increase in world population will take place in LDCs, where 1 million people are added every 4 days (Figure 1-5). By 2010 the combined populations of Asia and Africa are projected to be 5.3 billion—almost as many as now live on the entire planet.

The growing gap since 1960 between rich and poor in per capita income has widened further since 1980 (Figure 1-6). The rich have grown much richer while the poor have stayed poor or grown even poorer. Today, one person in five lives in luxury while the next three "get by" and the fifth struggles to survive on less than $1 a day (Spotlight, p. 7).

1-3 Resources

A **resource** is anything we get from the living or nonliving environment to meet our needs and wants. We usually define resources in terms of humans, but all forms of life need resources for survival and good health. Some resources, such as solar energy, fresh air, fresh surface water, fertile soil, and wild edible plants, are directly available for use by us and other organisms. Most human resources, such as petroleum (oil), iron, groundwater (water occurring underground), and modern crops, aren't directly available, and their supplies are limited. They become useful to us only with some effort and technological ingenuity. Petroleum, for example, was a mysterious fluid until we learned how to find it, extract it, and refine it into gasoline, heating oil, and other products that could be sold at affordable prices. We typically classify resources as renewable, potentially renewable, or nonrenewable (Figure 1-7).

NONRENEWABLE RESOURCES Nonrenewable (or **exhaustible**) **resources** exist in fixed quantities in the earth's crust. On a time scale of millions to billions of years, such resources can be renewed by geological processes. However, on the much shorter human time scale of hundreds to thousands of years, these resources can be depleted much faster than they can be formed.

These exhaustible resources include *energy resources* (coal, oil, natural gas, uranium), *metallic mineral resources* (iron, copper, aluminum), and *nonmetallic*

Q: Where does everything that supports your life come from?

The Desperately Poor

SPOTLIGHT

Because of population growth and the wealth gap,

- One person out of five is hungry or malnourished and lacks clean drinking water, decent housing, and adequate health care.
- One person out of three lacks enough fuel to keep warm and to cook food.
- One adult out of four cannot read or write.
- More than half of humanity lacks sanitary toilets.
- One person out of every five on Earth is desperately poor—unable to grow or buy enough food to stay healthy or to work.

Life for the desperately poor is a harsh daily struggle for survival. Poor parents—some with seven to nine children—are lucky to have an annual income of $300 (82¢ a day). Having many children makes good sense to most poor parents because their children represent a form of economic security by helping them grow food or tend livestock, or by holding a job or begging in the streets. And the two or three offspring who live to adulthood can help their parents survive in the latter's old age (forties or fifties).

Each year an estimated 40 million of the desperately poor die from malnutrition (lack of protein and other nutrients needed for good health) or related diseases and from contaminated drinking water. This death toll of 110,000 people *per day* (over 40 million annually) is equivalent to 275 planes, each carrying 400 passengers, crashing every day. Half of those who die are children under 5, with most dying from diarrhea and measles because they are weakened by malnutrition. Even a lower estimated annual death toll of about 20 million people is appalling. *The dying are unique human beings, not mere numbers or things.*

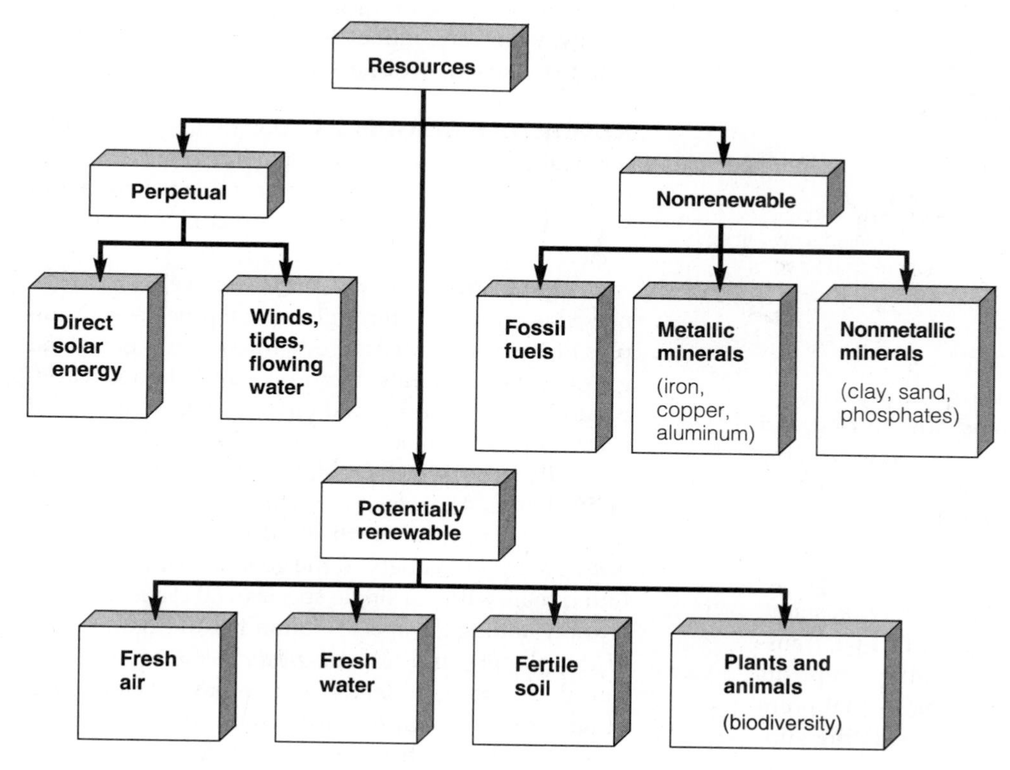

Figure 1-7 Major types of material resources. This scheme isn't fixed; potentially renewable resources can become nonrenewable resources if used for a prolonged period at a faster rate than they are renewed by natural processes.

A: The sun and the earth (Earth capital)

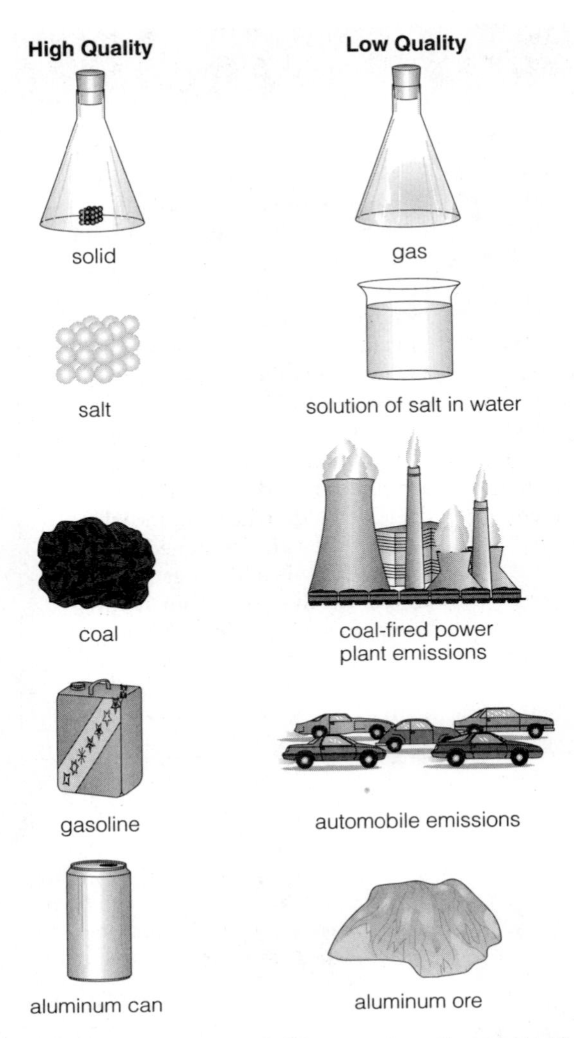

Figure 1-8 Examples of differences in matter quality. High-quality matter is fairly easy to extract and is concentrated. Low-quality matter is more difficult to extract and is more dispersed than high-quality matter.

mineral resources (salt, gypsum, clay, sand, phosphates, water, soil) (Figure 1-7). We know how to find and extract more than 100 nonrenewable minerals from the earth's crust. We convert these raw materials into many everyday items we use, and then we discard, burn, reuse, or recycle them.

We never completely run out of any nonrenewable mineral. However, a mineral becomes *economically depleted* when the costs of finding, extracting, transporting, and processing the remaining deposits exceeds the amount earned from them. At that point we have five choices: recycle or reuse existing supplies, waste less, use less, try to develop a substitute, or do without and wait millions of years for more to be produced.

Some nonrenewable material resources, such as copper and aluminum, can be recycled or reused to extend supplies. **Recycling** involves collecting and reprocessing a resource into new products. For example, aluminum cans can be collected, melted, and made into new beverage cans or other aluminum products. And glass bottles can be crushed and melted to make new bottles or other glass items. **Reuse** involves using a resource over and over in the same form. For example, glass bottles can be collected, washed, and refilled many times.

Nonrenewable energy resources—such as coal, oil, and natural gas—can't be recycled or reused. Once burned, the useful energy in these fossil fuels is gone, leaving behind waste heat and polluting exhaust gases.

We can classify resources in terms of **matter quality**—a measure of how useful a matter resource is, based on its availability and concentration. **High-quality matter** is organized and concentrated and is usually found near the earth's surface. It has great potential for use as a matter resource. **Low-quality matter** is disorganized, dilute, or dispersed, and it is often found deep underground or dispersed in the ocean or in the atmosphere. It usually has little potential for use as a matter resource (Figure 1-8). An aluminum can, for example, is a more concentrated, higher-quality form of aluminum than aluminum ore containing the same amount of aluminum. That's why it takes less energy, water, and money to recycle an aluminum can than to make a new can from aluminum ore.

RENEWABLE RESOURCES Solar energy is called a **renewable resource** because on a human time scale it is essentially inexhaustible. It is expected to last at least 4 billion years while the sun completes its life cycle.

A **potentially renewable resource*** can be renewed fairly rapidly through natural processes. Examples of such resources include forest trees, grassland grasses, wild animals, fresh lake and stream water, groundwater, fresh air, and fertile soil. One important potentially renewable resource is **biological diversity**, or **biodiversity**, which consists of the numerous forms of life that can best survive the multitude of conditions currently found on Earth. Biodiversity includes **(1)** *genetic diversity* (variety in the genetic makeup among individuals within a single species), **(2)** *species diversity* (variety among the species found in different habitats of the planet), and **(3)** *ecological diversity* (variety of forests, deserts, grasslands, streams, lakes, oceans, and other biological communities that interact with one another and with their nonliving environments). This

*Most sources use the term *renewable resource*. The word *potentially* is added here to emphasize that these resources can be depleted if we use them faster than natural processes renew them.

Q: How many of the world's people live in absolute poverty?

rich variety of genes, species, and ecosystems gives us food, wood, fibers, energy, raw materials, industrial chemicals, and medicines—all of which pour hundreds of billions of dollars yearly into the world economy. Earth's vast inventory of life forms and ecosystems also provides free recycling and purification services and natural pest control. Every species found on the earth today contains stored genetic information that is the raw material for future adaptations. Biodiversity is nature's "insurance policy" against disasters.

Potentially renewable resources, however, can be depleted. The highest rate at which a potentially renewable resource can be used without reducing its available supply is called its **sustainable yield**. If a resource's natural replacement rate is exceeded, the available supply begins to shrink—a process known as **environmental degradation**.

The following types of environmental degradation can change potentially renewable resources into nonrenewable or unusable resources:

- *Covering productive land with water, concrete, asphalt, or buildings so that plant growth declines and wildlife habitats are lost.*
- *Cultivating land without proper soil management, causing soil erosion and depletion of plant nutrients.* Topsoil is now eroding faster than it forms on about 33% of the world's cropland—a loss of about 23 billion metric tons (25 billion tons) per year.
- *Irrigating cropland without good drainage, causing salinization or waterlogging.* Salt buildup has cut yields on one-fourth of all irrigated cropland, and waterlogging has reduced productivity on at least one-tenth of such croplands.
- *Taking fresh water from underground sources (aquifers) and from streams and lakes faster than it is replaced by natural processes.* In the United States, one-fourth of the groundwater withdrawn each year is not replenished.
- *Destroying wetlands and coral reefs.* Between 25% and 50% of the world's wetlands have been drained, built on, or seriously polluted. The United States has lost 55% of its wetlands and loses another 150,000 hectares (371,000 acres) each year. Coral reefs are being destroyed or damaged in 93 of the 109 major locations.
- *Cutting trees from large areas without adequate replanting (deforestation).* Almost half of the world's tropical forests have been cleared. If current deforestation rates continue, within 30 to 50 years there may be little of these forests left. In MDCs many of the remaining diverse, old-growth forests are being cleared and replaced with single-species tree farms or with often much less diverse second-growth forests. Such practices reduce wildlife habitats and forms of wildlife that contribute to the planet's biodiversity.
- *Overgrazing of grassland by livestock, which converts productive grasslands into unproductive land (desertification).* Each year almost 60,000 square kilometers (23,000 square miles) of new desert are formed, mostly because of overgrazing.
- *Eliminating or decimating wild species (biodiversity) through destruction of habitats, commercial hunting, pest control, and pollution.* Each year thousands of wildlife species become extinct, mostly because of human activities. If habitat destruction continues at current rates, as many as 1.5 million species could disappear over the next 25 years—a drastic loss in vital Earth capital.
- *Polluting renewable air, water, and soil so that they become unusable.*

The greatest danger from high levels of resource consumption may not be the exhaustion of resources, but rather the damage that their extraction and processing do to the environment. The mining, processing, and use of mineral resources require enormous amounts of energy and often cause land disturbance, erosion, and pollution of air and water (Figure 1-9). Environmentalists and some economists argue that until we include these harmful external costs in the market prices of goods, we will have little incentive to reduce environmental damage from resource use (Solutions, p. 5).

1-4 Pollution

WHAT IS POLLUTION? Anything added to air, water, soil, or food that threatens the health, survival capability, or activities of humans or other living organisms is called **pollution**. Most pollutants are solid, liquid, or gaseous by-products or wastes produced when a resource is extracted, processed, made into products, or used (Figure 1-9). Pollution can also take the form of unwanted or harmful energy emissions, such as excessive heat, noise, or radiation.

A major problem is that people can disagree about whether something is a pollutant and about acceptable levels of pollution, especially if they must choose between pollution control and their jobs. As the

A: At least 1.2 billion—about one person in every five on Earth

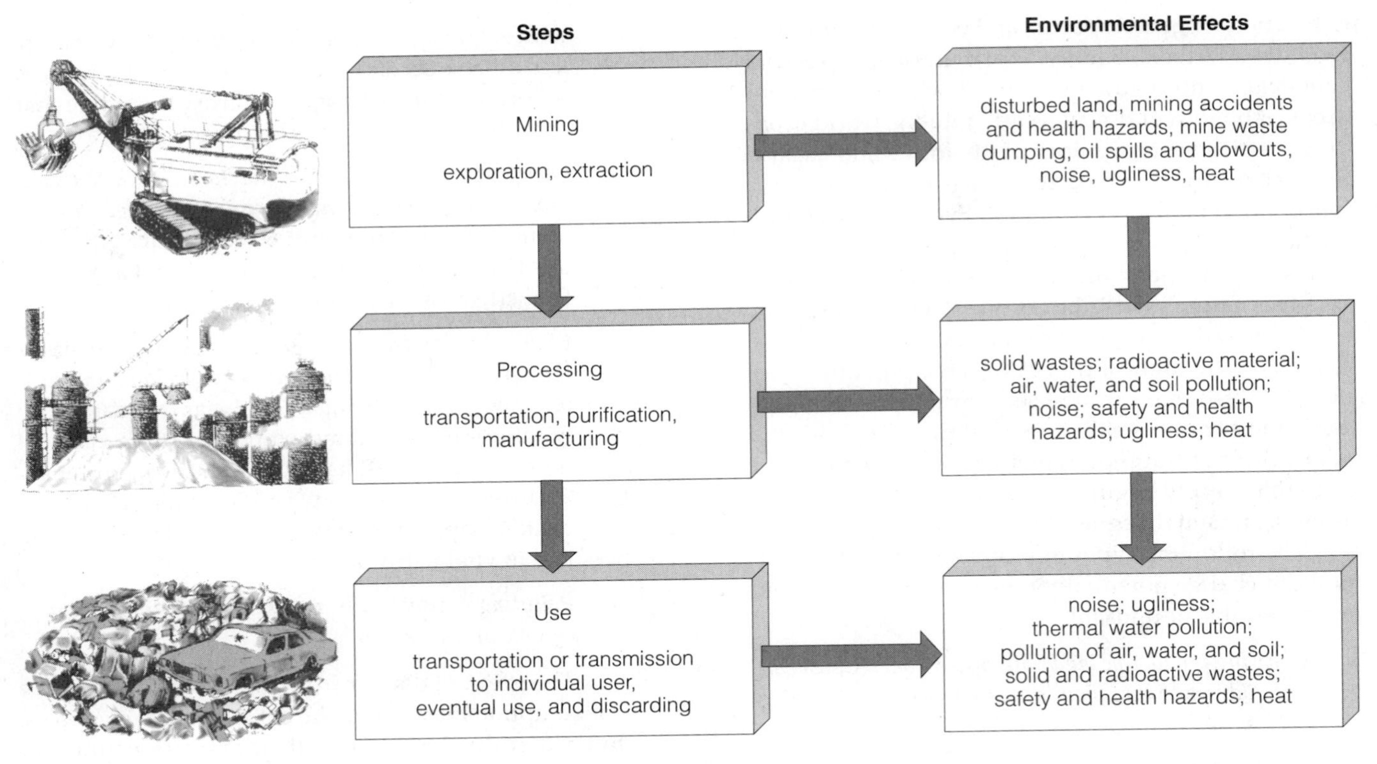

Figure 1-9 Some harmful environmental effects of resource extraction, processing, and use. The energy used to carry out each step causes additional pollution and environmental degradation.

philosopher Hegel pointed out, tragedy is not the conflict between right and wrong, but the conflict between right and right.

SOURCES OF POLLUTION Pollutants can enter the environment naturally (for example, from volcanic eruptions) or through human activities (for example, from burning coal). Most natural pollution is dispersed over a large area and is diluted or broken down to harmless levels by natural processes. By contrast, most serious pollution from human activities occurs in or near urban and industrial areas, where pollutants are concentrated in small volumes of air, water, and soil. Industrialized agriculture is also a major source of pollution.

Some pollutants contaminate the areas in which they are produced. Others are carried by winds or flowing water to other areas. Pollution does not respect state or national boundaries.

EFFECTS OF POLLUTION Effects of pollutants include **(1)** disruption of life-support systems for humans and other species, **(2)** damage to wildlife, **(3)** damage to human health, **(4)** damage to property, and **(5)** nuisance effects such as noise and unpleasant smells, tastes, and sights.

Three factors determine how severe the effects of a pollutant will be. One is its *chemical nature*—how active and harmful it is to living organisms; another is its *concentration*—the amount per unit volume of air, water, soil, or body weight. One way to lower the concentration of a pollutant is to dilute it in a large volume of air or water. Before we started overwhelming the air and waterways with pollutants, dilution was *the* solution to pollution. Now it is only a partial solution.

The third factor is a pollutant's *persistence*—how long it stays in the air, water, soil, or body. *Degradable* (or *nonpersistent*) *pollutants* are broken down completely or reduced to acceptable levels by natural physical, chemical, and biological processes. Pollutants broken down by living organisms (usually by specialized bacteria) are called *biodegradable pollutants*. Human sewage in a river, for example, is biodegraded fairly quickly by bacteria if the sewage is not added faster than it can be broken down.

Unfortunately, many of the substances we introduce into the environment take decades or longer to degrade. Examples of these *slowly degradable* or *persistent pollutants* include the insecticide DDT, most plastics, and chlorofluorocarbons (CFCs), which are used as coolants in refrigerators and air conditioners, as spray propellants (in some countries), and as foaming agents for making some plastics.

Q: How many children under age 5 die each day in poor countries of causes that could be prevented?

Nondegradable pollutants cannot be broken down by natural processes. Examples include the toxic elements lead and mercury. The best ways to deal with nondegradable pollutants (and slowly degradable hazardous pollutants) are to not release them into the environment in the first place, to recycle them, or to remove them from contaminated air, water, or soil (an expensive process).

We know little about the possible harmful effects of most of the 70,000 synthetic chemicals now in commercial use, mostly because it is quite difficult, time-consuming, and expensive to obtain this knowledge. According to the National Academy of Sciences, only about 10% of the 70,000 chemicals currently in commercial use have been thoroughly screened for toxicity, and only 2% have been adequately tested to determine whether they are carcinogens (cause cancer), teratogens (cause birth defects), or mutagens (cause harmful genetic changes that can be inherited in test organisms). Furthermore, each year about 1,000 new chemicals are introduced into the marketplace, often with too little knowledge about their potentially harmful effects. Even if we determine the biggest risks associated with a particular chemical, we know little about its possible interactions with other chemicals or about the effects of such interactions on human health and ecosystems.

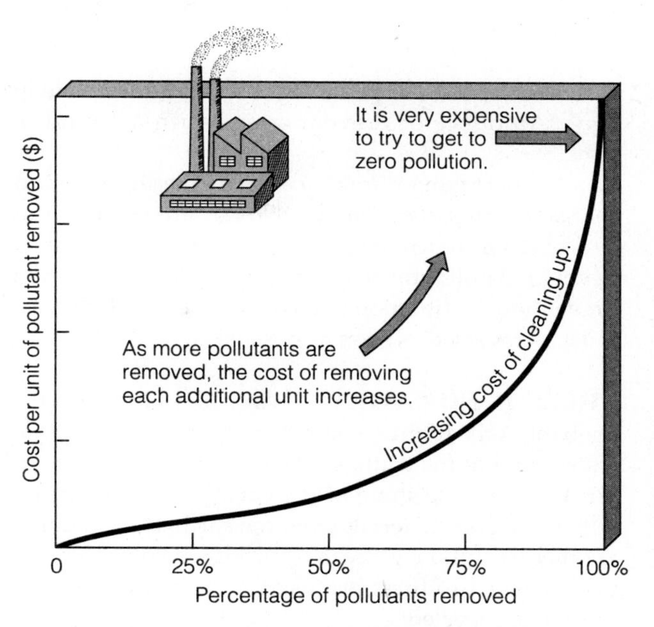

Figure 1-10 The cost of removing each additional unit of a pollutant rises exponentially, which explains why it is usually cheaper to prevent pollution than to clean it up.

DEALING WITH POLLUTION: PREVENTION AND CLEANUP **Pollution prevention**, or **input pollution control**, reduces or eliminates the input of pollutants and wastes into the environment. Pollution can be prevented, or at least reduced, by the three R's of resource use: *reduce, reuse, recycle.*

Pollution cleanup, or **output pollution control**, deals with pollutants after they have been produced. Relying primarily on pollution cleanup causes several problems. One is that as long as population and consumption levels continue to grow, cleanup is only modestly effective. For example, although adding catalytic converters to cars has reduced air pollution, increases in the number of cars and in the total distance each travels have reduced the effectiveness of this cleanup approach.

Another problem is that cleanup often removes a pollutant from one part of the environment, only to cause pollution in another part. We can collect garbage, but the garbage is then burned (perhaps causing air pollution and leaving a toxic ash that must be put somewhere); dumped into streams, lakes, and oceans (perhaps causing water pollution); buried (perhaps causing soil and groundwater pollution); recycled; or reused.

Should our goal be zero pollution? Ideally, yes; in the real world, not necessarily. First, nature can handle some of our biodegradable wastes, so long as we don't destroy, degrade, or overload these natural processes. But toxic products that cannot be degraded by natural processes or that break down very slowly in the environment should not be produced and released into the environment or should be used only in small amounts regulated by special permit.

Second, so long as we continue to rely on pollution control, we can't afford zero pollution. We can remove a certain percentage of the pollutants in air, water, or soil, but when we remove more, the cleanup cost per unit of pollutant removed rises sharply (Figure 1-10). Beyond a certain point, the cleanup costs will exceed the costs resulting from the harmful effects of pollution. If we don't go far enough, dealing with the harmful effects of pollution will cost more than pollution reduction.

Environmentalists argue that effective pollution prevention requires the assumption that any waste or pollutant is potentially harmful until shown otherwise. This *precautionary principle* is the inverse of the waste management approach, in which wastes and pollutants are assumed to be benign until shown to be harmful. Proponents of the waste management approach say we can't afford to place the burden of proof on manufacturers of goods; environmentalists say we can't afford not to reduce pollution and waste by internalizing these harmful external costs, which we end up paying for indirectly (Solutions, p. 5).

Both pollution prevention and pollution cleanup are needed, but environmentalists urge us to emphasize prevention because it works better and is cheaper

A: At least 40,000 per day (1,700 during your lunch hour)

than cleanup. As Benjamin Franklin reminded us long ago, "An ounce of prevention is worth a pound of cure."

An increasing number of businesses have found that *pollution prevention pays*. So far, however, about 99% of environmental spending in the United States is devoted to pollution cleanup and only 1% to pollution prevention—a situation that environmentalists believe must be reversed as soon as possible.

THERE IS NO "AWAY" You, like most people, probably talk about consuming or using up material resources, but the truth is that we don't consume matter: We only use some of the earth's resources for a while. We take materials from the earth, carry them to another part of the globe, and process them into products. These products are used and then discarded, reused, or recycled.

We may change various elements and compounds from one physical or chemical form to another, but in no physical and chemical changes can we create or destroy any of the atoms involved. All we can do is rearrange them into different spatial patterns (physical changes) or different combinations (chemical changes). This fact, based on many thousands of measurements of matter undergoing physical and chemical changes, is known as the **law of conservation of matter**.

The law of conservation of matter means that there really is no "away" to throw things to. *Everything we think we have thrown away is still here with us in one form or another.*

The law of conservation of matter means that we will always be faced with the problem of what to do with some quantity of chemical wastes. By placing much greater emphasis on pollution prevention and waste reduction, however, we can greatly reduce the amount of chemical wastes we add to the environment.

Table 1-1 Greatest Ecological and Health Risks*

High-risk Ecological Problems
Global climate change
Stratospheric ozone depletion
Wildlife habitat alteration and destruction
Species extinction and loss of biodiversity
Medium-risk Ecological Problems
Acid deposition
Pesticides
Airborne toxic chemicals
Toxic chemicals, nutrients, and sediment in surface waters
Low-risk Ecological Problems
Oil spills
Groundwater pollution
Radioactive isotopes
Acid runoff to surface waters
Thermal pollution
High-risk Health Problems
Indoor air pollution
Outdoor air pollution
Worker exposure to industrial or farm chemicals
Pollutants in drinking water
Pesticide residues on food
Toxic chemicals in consumer products

*Items in each category are not listed in rank order. Data from Science Advisory Board, *Reducing Risks*, Washington, D.C., Environmental Protection Agency, 1990.

1-5 Environmental Problems: Causes and Connections

ROOT CAUSES OF ENVIRONMENTAL PROBLEMS The first step in dealing with the interconnected array of environmental problems we face is to identify which ones pose the greatest threats both to life-support systems for us and other species and to human health. Table 1-1 ranks the most serious ecological and health risks identified by a panel of scientists acting as advisers to the U.S. Environmental Protection Agency.

The next step in dealing with environmental problems is to identify their underlying causes, which include the following:

- *Overpopulation.*
- *Overconsumption of resources*, especially by the affluent.
- *Poverty.*
- *Resource waste.*
- *Widespread use of environmentally damaging fossil fuels* (especially oil and coal).
- *Loss of biodiversity through oversimplification of the earth's life-support systems.*
- *Overuse and degradation of common-property resources, which are owned by none and available to*

Q: What percentage of environmental spending in the United States is devoted to preventing pollution?

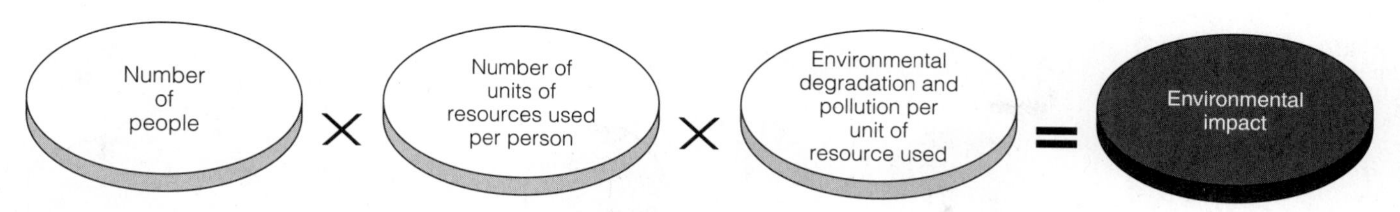

Figure 1-11 Simplified model of how three factors—population (P), affluence (A), and technology (T)—affect the environmental impact (environmental degradation and pollution) of population.

all. Examples include clean air, fish in the open ocean, migratory birds, Antarctica, gases of the lower atmosphere, and the ozone content of the upper atmosphere. In 1968 biologist Garrett Hardin called this *the tragedy of the commons*. It happens because each user reasons, "If I don't use this resource, someone else will. The little bit I use or pollute is not enough to matter." When there are only a few users, this logic works. However, the cumulative effects of many people trying to exploit a common-property resource eventually exhaust or ruin it, so that no one can benefit from it. Therein lies the tragedy.

- *Failure to encourage Earth-sustaining forms of economic development and discourage Earth-degrading forms of economic growth* (Solutions, p. 5).
- *Our urge to dominate and control nature for our use.*

CONNECTIONS AMONG ROOT CAUSES OF ENVIRONMENTAL PROBLEMS Once we have identified the causes of our problems, the next step is to understand how they are connected to one another. The three-factor model in Figure 1-11 is a good starting point.

According to this model, total environmental degradation and pollution—that is, the environmental impact—in a given area depends on three factors: **(1)** the number of people (population size, P), **(2)** the average number of units of resources each person uses (consumption per capita or affluence, A), and **(3)** the amount of environmental degradation and pollution produced for each unit of resource used (the environmental destructiveness of the technologies used to provide and use resources, T). This model, developed in the early 1970s by biologist Paul Ehrlich and physicist John Holdren, can be summarized in simplified form as Impact = Population × Affluence × Technology, or $I = P \times A \times T$.

Overpopulation occurs when too many people deplete the resources that support life and economies, and when people introduce into the environment more wastes than it can handle. Overpopulation happens when people exceed an area's **carrying capacity**: the number of people an area can support given its resource base and the way those resources are used. According to the developers of this model, the three factors in Figure 1-11 can interact to produce two types of overpopulation (Figure 1-12).

People overpopulation exists when there are more people than the available supplies of food, water, and other important resources can support at a minimal level. In this case, population size and the resulting degradation of potentially renewable resources (as the poor struggle to stay alive) tend to be the key factors determining total environmental impact.

Consumption overpopulation, or **overconsumption**, exists in MDCs, in which only one-fifth of the world's people use resources at such a high rate that significant pollution, environmental degradation, and resource depletion occur. With this type of overpopulation, high rates of per capita resource use (and the resulting high levels of pollution and environmental degradation per person) are the key factors determining overall environmental impact. For example, the average U.S. citizen consumes 50 times more resources than the average citizen of India (and 100 times more than the average person in some LDCs). *Thus poor parents in an LDC would need to have 100–200 children to produce the same lifetime environmental impact as two children in a typical U.S. family.*

We know from studying other species that when a population exceeds or *overshoots* the carrying capacity of its environment, it suffers a *dieback* that reduces its population to a sustainable size. How long will we be able to continue our exponential growth in population (Figure 1-1) and resource use (Figure 1-2) without suffering overshoot and dieback? No one knows, but warning signals from the earth are forcing us to consider the question seriously.

Some analysts (mostly economists), however, contend that the world is not overpopulated and that more people are our most important resource (as consumers to fuel continued economic growth and as sources of technological ingenuity). They believe that technological advances will allow us both to clean up pollution to acceptable levels and to find substitutes for any resources that become scarce. They also accuse environmentalists of overstating the seriousness of the problems we face.

A: 1%; the other 99% is spent on pollution cleanup

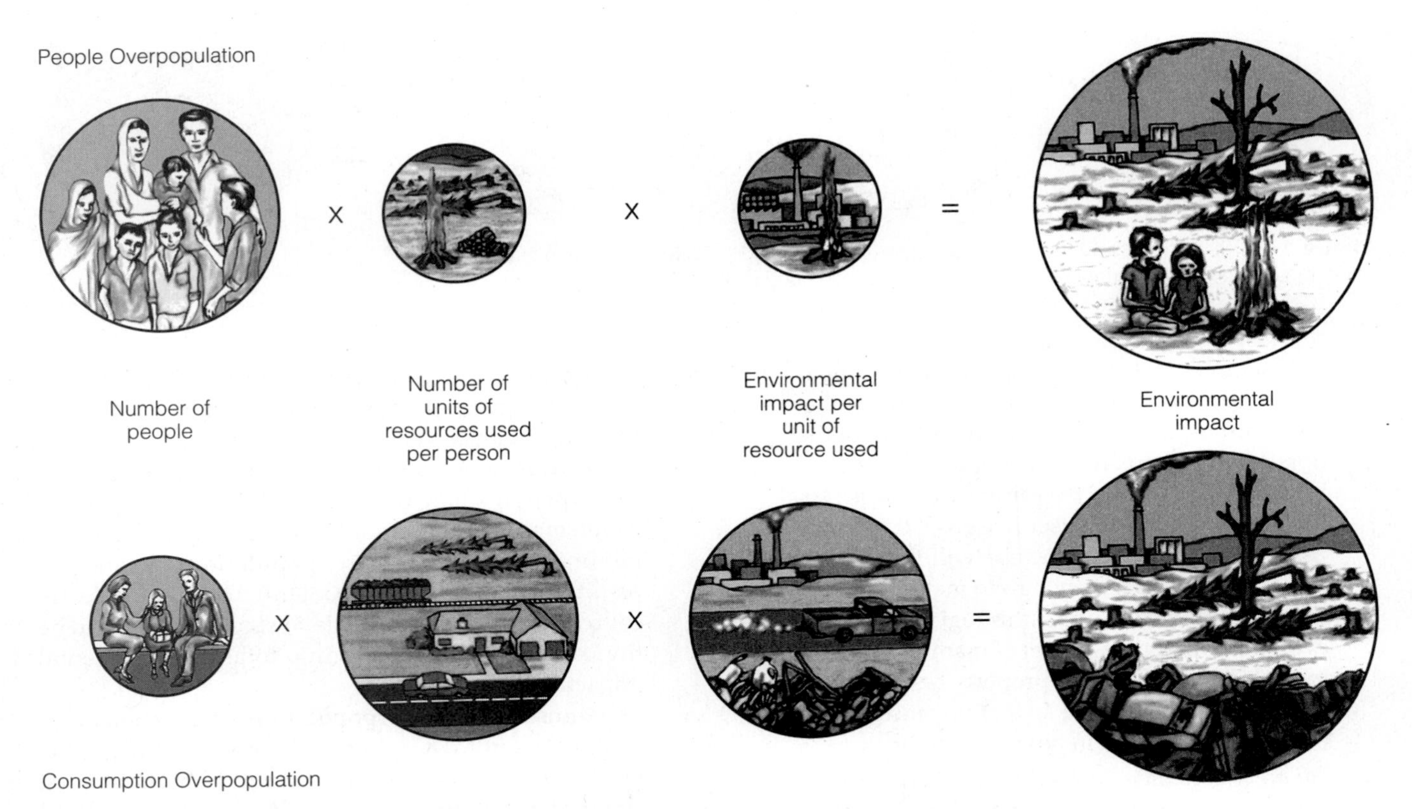

Figure 1-12 Two types of overpopulation based on the relative importance of the factors in the model shown in Figure 1-11. Circle size shows relative importance of each factor.

LIMITATIONS AND MISUSE OF SCIENCE Some people misunderstand the nature and limitations of science, and some may deliberately misuse it. Those who say that something has or has not been "scientifically proven" imply falsely that science yields absolute proof or certainty. *Scientists can disprove things, but they can't establish absolute proof or truth.*

Instead of certainty, science gives us information in the following form: If we do such-and-such (say, add certain chemicals to the atmosphere at particular rates), there is a certain chance (high, moderate, or low) that we will cause various effects (such as changing the climate or depleting ozone in the stratosphere). The more complex the system or problem being studied, the less certain the hypotheses, models, and theories used to explain that system or problem.

Science advances by debate, argument, speculation, and controversy concerning the findings of researchers. Disputes among scientists over the validity of data, hypotheses, and models (tentative ideas still being rigorously evaluated and tested) are what the media often report. Such disagreements make juicier stories, but what's really important is the *consensus* among scientists about various experiments, theories, models, and laws. This substantial agreement—which determines the real knowledge of science—rarely gets reported, giving the public a false idea of the nature of science and of scientific knowledge.

Most environmental problems involve such complex mixtures of data, lack of data, hypotheses, and theories in the physical and social sciences that we don't have enough information to understand them very well. This allows advocates of any proposed cause of action (or inaction) on an environmental problem to support their beliefs. This can cause confusion and controversy among the general public and elected officials and can lead to misguided government regulation. It can also lead to the *paralysis-by-analysis* trap that insists that we fully understand a problem before taking any action—an impossible dream because of the inherent limitations of science and the complexities of environmental problems.

Because environmental problems won't go away, at some point we must evaluate the available information and make political and economic decisions about what to do (or not do), often based primarily on gut feelings, intuition, common sense (which proponents of every viewpoint claim they are using), and values. People with different values and views about how the world works can take the same information, come to completely different conclusions, and still be logically consistent. There are no easy answers.

Q: What percentage of U.S. municipal solid waste could be recycled, composted, or reused?

What It Means to Live in a Throwaway Society

CONNECTIONS

With only 4.7% of the world's population, the United States produces 33% of the world's *municipal solid waste* or *garbage*. U.S. consumers throw away:

- Enough aluminum to rebuild the country's entire commercial airline fleet every three months.
- Enough glass bottles to fill the two 412-meter-high (1,350-foot) towers of the New York World Trade Center every two weeks.
- Enough tires each year to encircle the earth almost three times.
- Enough disposable plates and cups each year to serve everyone in the world six meals.
- About 18 billion disposable diapers per year, which if linked end to end would reach to the moon and back seven times.
- About 2 billion disposable razors, 1.6 billion throwaway ballpoint pens, and 500 million disposable cigarette lighters each year.
- About 2.5 million nonreturnable plastic bottles each hour.
- Some 14 billion catalogs (an average of 54 per person) and 38 billion pieces of junk mail per year.

And this is only part of the 1.5% of solid waste produced in the United States that we call garbage!

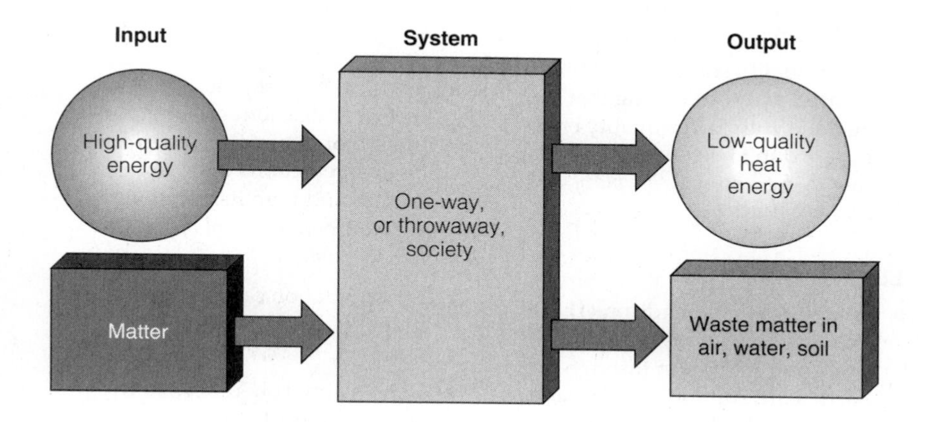

Figure 1-13 The high-waste or *throwaway society* of most MDCs is based on increasing economic growth by maximizing the rates of energy and matter flow, thereby rapidly converting the world's high-quality matter and energy resources into waste, pollution, and low-quality heat. To achieve long-term sustainability, such a society must have an infinite supply of resources (including energy), the ability to find a substitute for any depleted resource, and an infinite capacity of the environment to renew renewable resources and dilute or degrade wastes and pollution.

1-6 Working with the Earth: The Low-Waste Society

Most MDCs are largely *high-waste* or *throwaway societies* that support ever-increasing economic growth by increasing the flow of matter and energy resources (Earth capital) through their economies (Figure 1-13 and Connections, above). Most of the wastes and pollutants produced end up in the environment.

In the United States, 83% of the *municipal solid waste* or *garbage* from homes and businesses is hauled away and either dumped (66%) or burned (17%). However, this garbage—which in 1993 would fill a bumper-to-bumper convoy of garbage trucks encircling the earth almost six times—makes up only 1.5% of the solid waste produced in the United States.

The remaining 98.5% of the solid waste in the United States—the world's largest producer of solid waste—comes from mining, oil and natural gas production, agriculture, and industrial activities. Although individuals don't generate this waste directly, they are responsible indirectly through the products they consume.

The United States is also the world's largest producer of hazardous waste. However, because only 6% of the total is legally defined as hazardous waste by Congress and is thus subject to government regulation, *94% of the country's hazardous waste is not regulated by hazardous waste laws*. Environmentalists urge that all currently excluded categories be designated hazardous waste—a decision that would quickly shift the emphasis from waste management and pollution control to waste reduction and pollution prevention and would in the long run save lots of money.

A: 60–80%

Principles for Understanding and Working with the Earth

SPOTLIGHT

Nature of Science

- Science is an attempt to discover order in nature and then use that knowledge to describe, explain, and predict what happens in nature.
- Scientists don't establish absolute proof or truth. Scientific laws, hypotheses, and theories are based on statistical probabilities, not on certainties, and are constantly being tested and challenged.

Matter

- Matter cannot be created or destroyed; it can only be changed from one form to another. Everything we think we have thrown away is still with us in one form or another; there is no away (*law of conservation of matter*).
- Organized and concentrated matter is high-quality matter that can usually be converted into useful resources at an affordable cost; disorganized and dispersed matter is low-quality matter that often costs too much to convert to a useful resource (*concept of matter quality*).

Energy

- Energy cannot be created or destroyed; it can only be changed from one form to another. We can't get energy for nothing; in terms of energy quantity it takes energy to get energy (*first law of energy or thermodynamics, or law of conservation of energy*).
- Organized or concentrated energy is high-quality energy that can be used to do things; disorganized or dilute energy is low-quality energy that is not very useful (*concept of energy quality*).
- In any conversion of energy from one form to another, high-quality, useful energy is always degraded to lower-quality, less useful energy that can't be recycled to give high-quality energy; we can't break even in terms of energy quality (*second law of energy or thermodynamics*).
- High-quality energy should not be used to do something that can be done with lower-quality energy; we don't need to use a chain saw to cut butter (*principle of energy efficiency*).

Life

- Life on Earth depends on **(1)** the one-way flow of high-quality energy from the sun through Earth's life-support systems and eventually back into space as low-quality heat, **(2)** gravity, and **(3)** the recycling of vital chemicals by a combination of biological, geological, and chemical processes (*principle of energy flow, gravity, and matter recycling*).
- Each species and each individual organism can tolerate only a certain range of environmental conditions (*range-of-tolerance concept*).
- Too much or too little of a physical or chemical factor can limit or prevent the growth of a population in a particular place (*limiting factor principle*).
- Every species has a specific role to play in nature (*ecological-niche concept*).
- Average precipitation and temperature are the major factors determining whether a particular land area is a desert, grassland, or forest (*climate-biome principle*).
- The size, growth rate, age structure, and distribution for a population of a species is controlled by its interactions with other species and with its nonliving environment (*concept of population dynamics*).
- No population can keep growing indefinitely (*concept of carrying capacity*).
- Earth's atmosphere, hydrosphere, lithosphere (upper crust and mantle), and forms of life are continually changing in response to changes in solar input, heat flows from the earth's interior, movements of the earth's crust, other natural changes, and changes brought

According to the National Academy of Sciences, the best way to deal with 60–80% of the wastes we produce is a preventive approach with the following hierarchy of goals: **(1)** *reduce* waste and pollution by preventing its creation; **(2)** *reuse* as many things as possible; **(3)** *recycle and compost* as much waste as possible; **(4)** *detoxify* as much of any remaining hazardous wastes as possible; **(5)** *incinerate* in state-of-the-art and rigorously regulated incinerators any remaining waste that can't be reduced, reused, recycled, composted, or detoxified; and **(6)** *bury* what is left in state-of-the-art and carefully regulated landfills after the first five goals have been met.

To date, most of the wastes we produce are burned or buried—the reverse of what many prominent scientists say we should be doing. Burning and burying wastes are seen as last resorts because they waste potential resources. Moreover, according to the Naabout by humans and other living organisms.

- Over billions of years, changes in environmental conditions have led to development of a variety of species (species diversity), genetic variety within species (genetic diversity), and natural systems (ecosystem diversity) through a mixture of extinction and formation of new species (*concept of biodiversity*).
- All species eventually become extinct by disappearing or by evolving into one or more new species in response to environmental changes brought about by natural processes or by human action.
- Earth's life-support systems can take much stress and abuse, but there are limits.

Humans and Environment

- Our survival, life quality, and economies are totally dependent on the sun and the earth; Earth can get along without us, but we can't get along without the earth (*principle of Earth capital*).
- We are part of—not apart from—nature. The earth does not belong to us; we belong to the earth.
- We should try to understand and work with the rest of nature to sustain the ecological integrity, biodiversity, and adaptability of Earth's life-support systems for us and other species.
- We can learn a lot about how nature works, but nature is so incredibly complex and dynamic that such knowledge will always be quite limited.
- In nature we can never do just one thing; everything we do creates effects that are often unpredictable (*first law of human ecology*).
- Everything is connected to and intermingled with everything else; we are all in this together; we need to understand these connections and discover those that are most important for sustaining life on the earth (*concept of interdependence or connectedness*).
- Most resources are limited and should not be wasted; there is not always more and it is not all for us.
- Living off renewable solar energy and renewable matter resources (Earth income) is a sustainable human lifestyle; using renewable matter resources faster than they are replenished and living off nonrenewable matter and energy resources degrade and deplete Earth capital and are ultimately an unsustainable lifestyle.
- Locally available and renewable resources should be used where possible; renewable resources should be used no faster than they're replenished by natural processes (*concept of sustainable yield*).
- Increases in population, resource use, or both will eventually overwhelm attempts to control pollution and manage wastes.
- Pollution prevention and waste reduction are the best and cheapest ways to sustain the earth. The best way to reduce pollution and waste is to not produce so much (*principle of pollution prevention and waste reduction*).
- The best way to protect species and individuals of species is to protect the habitats and ecosystems in which they live.
- We should change Earth-degrading and Earth-depleting manufacturing processes, products, and businesses into Earth-sustaining ones by using economic incentives and penalties.
- The market price of a product should include all estimated present and future costs of any pollution, environmental degradation, or other harmful effects connected with it that are passed on to society, the environment, and future generations (*concept of full-cost pricing*).
- The condition in which we leave the earth should be as good as (or better than) we found it.

tional Academy of Sciences and the Environmental Protection Agency, even the best-designed waste incinerators release some toxic substances into the atmosphere and leave a toxic residue that must be disposed of—usually in landfills. And even the best-designed landfills eventually leak wastes into groundwater. Eventually we will also run out of affordable—and politically acceptable—sites for landfills and incinerators.

Scientific principles based on how nature works (Spotlight, above) indicate that the best long-term solution to our environmental and resource problems is to shift from a *high-waste society* (Figure 1-13) to a *low-waste society* (Figure 1-14, p. 18) over the next few decades. Making this shift begins by trying to reduce the waste production of individuals (Individuals Matter, p. 19) and by recognizing four things: **(1)** that we are part of—not apart from—the earth's dynamic web

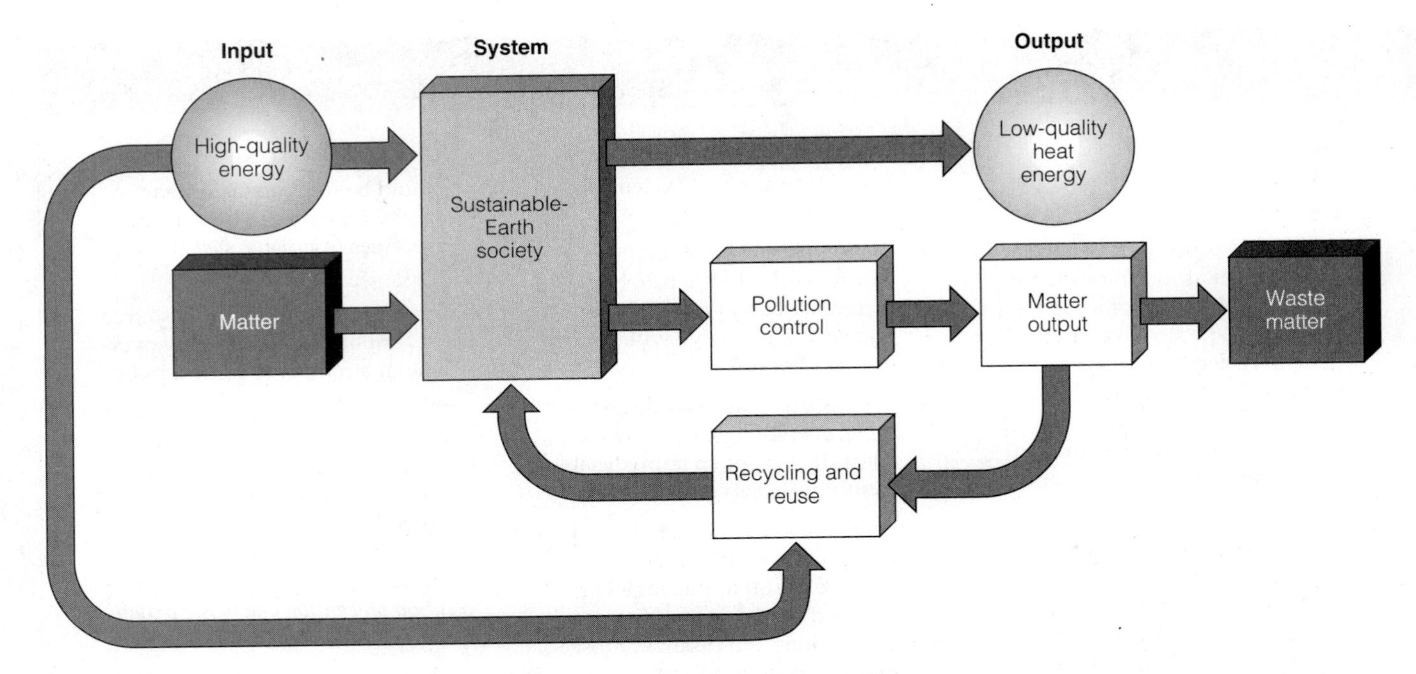

Figure 1-14 A *low-waste society* assumes that the availability of resources, opportunities for resource substitution, and the ability of the environment to accept wastes are limited and that many of these limits may be exceeded over the next 40–50 years. This type of society works with nature by reusing and recycling most nonrenewable matter resources, by using potentially renewable resources no faster than they are replenished, by using matter and energy resources efficiently, by reducing wasteful consumption, by emphasizing pollution prevention and waste reduction, and by controlling population growth.

of life; **(2)** that our survival, lifestyles, and economies are totally dependent on the sun and the earth; **(3)** that everything is connected to everything else; and **(4)** that there is no "away" for us to put our wastes.

This chapter has presented an overview of the major environmental issues and problems and their root causes. The rest of this book presents a more detailed analysis of these problems and presents some solutions proposed by various scientists and environmentalists.

The key to dealing with the problems we face is recognizing that *individuals matter*. Anthropologist Margaret Mead expressed our potential for change in these words: "Never doubt that a small group of thoughtful, committed citizens can change the world. Indeed it is the only thing that ever has."

Where there is no dream, the people perish.

PROVERBS 29:18

CRITICAL THINKING

1. If there is no "away" for us to put our wastes, why isn't the world filled with waste matter?
2. Do you favor internalizing the external costs of pollution and unnecessary resource waste? Explain your answer. How might it affect your lifestyle? Wildlife?
3. Do you favor instituting policies designed to reduce population growth and to stabilize both the size of the world's population and the size of the U.S. population as soon as possible? Explain. What policies do you believe should be implemented?
4. Keep a list for a week of the things you throw away. Roughly what percentage are materials that could be recycled or reused? What percentage of the items could you have done without?
5. Would you oppose having a hazardous-waste landfill, a treatment plant, a deep disposal well, or an incinerator in your community? Explain. If you oppose these disposal facilities, how should the hazardous waste generated in your community and your state be managed?
6. Only about 6% of the hazardous waste produced in the United States is subject to government regulation. Do you believe that all hazardous waste produced should be subject to government regulation? Explain. What are the possible long- and short-term effects on the economy of regulating all hazardous wastes? On your lifestyle? On any children you might have?

Q: What are the worst ways to deal with solid waste?

INDIVIDUALS MATTER

What *You* Can Do to Reduce Waste and Save Money

- *Buy less by asking yourself whether you really need a particular item.*
- *Buy things that are reusable or recyclable, and be sure to reuse and recycle them.*
- *Buy beverages in refillable glass containers instead of cans or throwaway bottles.*
- *Use plastic or metal lunch boxes and metal or plastic garbage containers without throwaway plastic liners.*
- *Carry sandwiches and store food in the refrigerator in reusable containers instead of wrapping them in aluminum foil or plastic wrap.*
- *Use rechargeable batteries and recycle them when their useful life is over.*
- *Carry groceries and other items in a reusable basket, a canvas or string bag, or a small cart.*
- *Use sponges and washable cloth napkins, dish towels, and handkerchiefs instead of paper ones.*
- *Use reusable or refillable plates, cups, and eating utensils whenever possible.*
- *Buy recycled goods and then recycle them.*
- *Reduce the amount of junk mail you get.* This can be accomplished at no charge by writing to Mail Preference Service, Direct Marketing Association, 11 West 42nd St., P.O. Box 3681, New York, NY 10163-3861, or by calling 212-768-7277, and asking that your name not be sold to large mailing-list companies. Of the junk mail you do receive, recycle as much of the paper as possible. You can also reduce junk mail by not accepting it and writing "Return to sender" on the envelope. This costs the mailer money.
- *Buy products in concentrated form whenever possible.*
- *Choose items that have the least packaging or, better yet, no packaging ("nude products").*
- *Compost your yard and food wastes, and lobby local officials to set up a community composting program.*
- *Use pesticides and other hazardous chemicals only when absolutely necessary, and in the smallest amount possible.*
- *Support legislation that would encourage pollution prevention and waste reduction.*

A: Burial and incineration

2 The Human Population

We shouldn't delude ourselves: The population explosion will come to an end before very long. The only remaining question is whether it will be halted through the humane method of birth control, or by nature wiping out our surplus.

PAUL EHRLICH

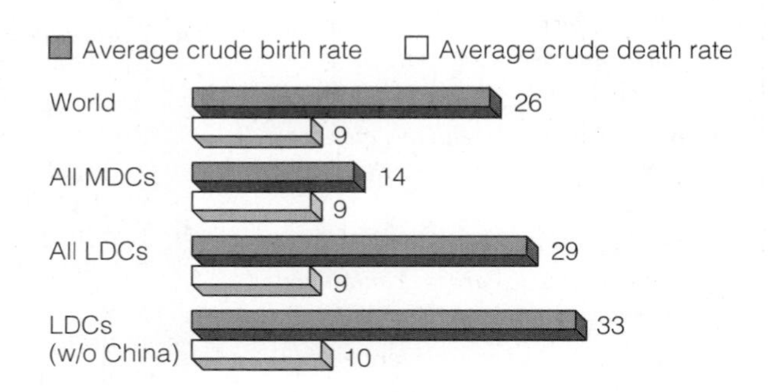

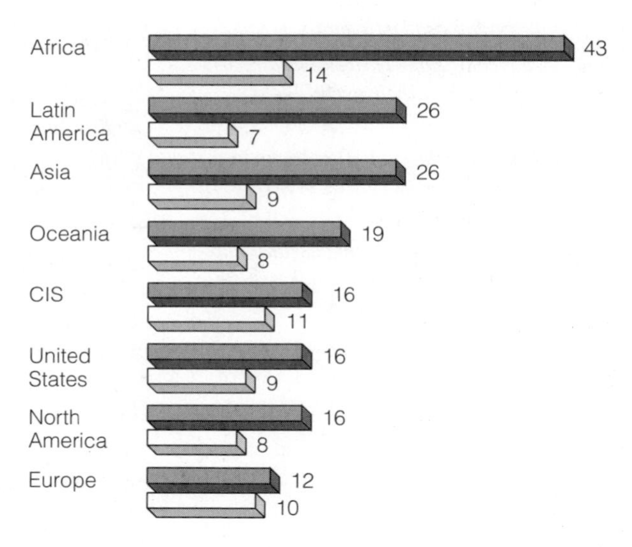

Figure 2-1 Average crude birth and death rates for various groups of countries in 1993. (Data from Population Reference Bureau)

2-1 Factors Affecting Human Population Size

BIRTH RATES AND DEATH RATES Populations grow or decline through the interplay of three factors: births, deaths, and migration. The **birth rate**, or **crude birth rate**, is the number of live births per 1,000 people in a population in a given year. The **death rate**, or **crude death rate**, is the number of deaths per 1,000 people in a population in a given year. Figure 2-1 shows the crude birth and death rates for various groups of countries in 1993. There are more births than deaths; every time your heart beats three more babies are added to the world's population—amounting to roughly a quarter of a million more people each day.

The world's annual population growth rate dropped 18% between 1965 and 1993, from 2% to 1.63%. This is good news, but during the same period the population base rose by 72%, from 3.2 billion to 5.5 billion. This 18% drop in the growth rate of population is akin to learning that the truck heading straight at you has slowed from 100 kilometers per hour to 82 kilometers per hour while its weight has increased by almost three-fourths. The current annual growth rate of 1.63% adds 90 million people per year—the equivalent of adding the population of Mexico each year and that of the United States about every three years.

In terms of sheer numbers of people, China and India dwarf all other countries, together making up 38% of the world's population. One person in five is Chinese, and 60% of the world's population is Asian. Even though the United States has the world's fourth largest population, it has only 4.7% of the world's people.

FERTILITY RATES Two types of fertility rates affect a country's population size and growth rate. The first type, **replacement-level fertility**, is the number of children a couple must bear to replace themselves. The actual average replacement-level fertility rate is slightly higher than two children per couple (2.1 in MDCs and as high as 2.5 in some LDCs), mostly because some female children die before reaching their reproductive years.

The second type of fertility rate, and the most useful measure for projecting future population change, is the **total fertility rate (TFR)**, an estimate of the aver-

Q: According to the EPA, how many landfills in the United States will eventually leak?

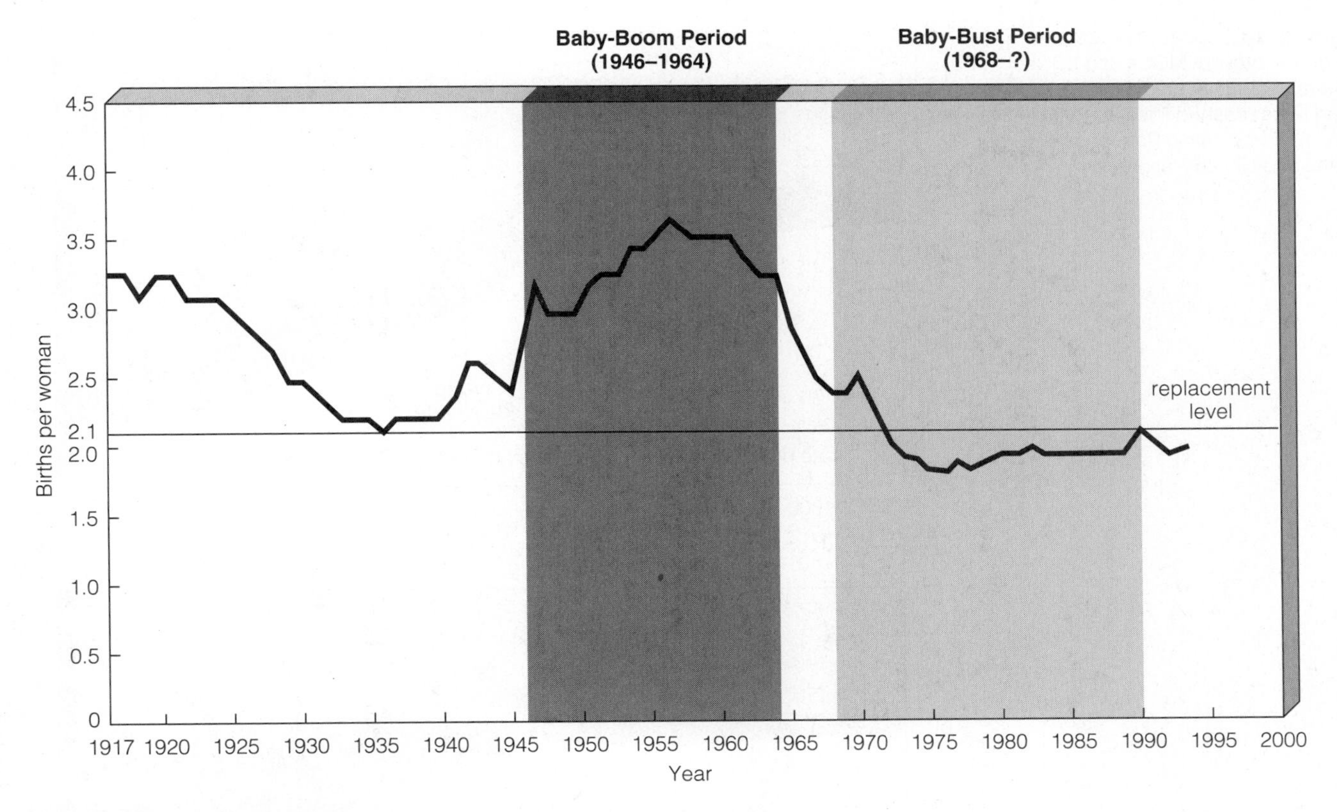

Figure 2-2 Total fertility rate for the United States between 1917 and 1993. (Data from Population Reference Bureau and U.S. Census Bureau)

age number of children a woman will have during her childbearing years. In 1993 the worldwide average TFR was 3.3 children per woman; it was 1.8 in MDCs (down from 2.5 in 1950) and 3.7 in LDCs (down from 6.5 in 1950). If the world's TFR remains at 3.3, its population will reach 694 billion by the year 2150—391 times the current population! Clearly this is not possible, but it does illustrate the enormous power of exponential population growth.

Population experts expect total fertility rates in MDCs to remain around 1.8 and those in LDCs to drop to around 2.3 by 2025. That is good news, but it will still lead to a projected world population of around 9 billion by 2025, with most of this growth taking place in LDCs (Figure 1-5, p. 6).

The population of the United States has grown from 4 million in 1790 to 258 million in 1993—a 64-fold increase—even though the country's total fertility rate has oscillated wildly (Figure 2-2). At the peak of the post–World War II baby boom (1946–64) in 1957, the total fertility rate reached 3.7 children per woman. Since then it has generally declined, remaining at or below replacement level since 1972. Various factors contributed to this decline in TFR:

- Widespread use of effective birth control methods.
- Availability of legal abortions.
- Social attitudes favoring smaller families.
- Greater social acceptance of childless couples.
- Increasing cost of raising a family. It will cost between $86,000 and $168,000 to raise to age 18 a child born in 1992.
- The rise in the average age at marriage between 1958 and 1991 from 20 to 24 for women and from 23 to 26 for men.
- More women working outside the home. In 1993 more than 70% of American women of childbearing age worked outside the home; their childbearing rate was one-third that of women not in the paid labor force.

The drop in the total fertility rate has led to a decline in the annual rate of population growth in the United States, but the country's population is still growing and is not even close to leveling off. In 1993 the U.S. population of 258 million grew by 1.2%—faster than that of any other MDC. This rate of growth added 3.1 million people, which is equivalent to adding another California every 10 years. The two main reasons for this growth are: **(1)** the large number of women (58 million) born during the baby-boom

A: All of them

Figure 2-3 Changes in crude birth and death rates for MDCs and LDCs between 1775 and 1993, and projected rates (dashed lines) to 2000. (Data from Population Reference Bureau and United Nations)

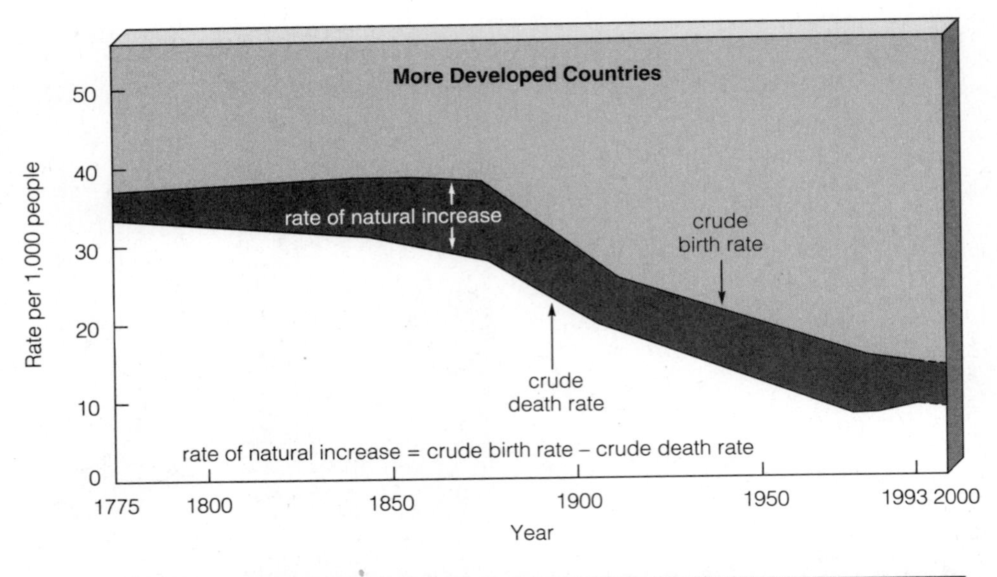

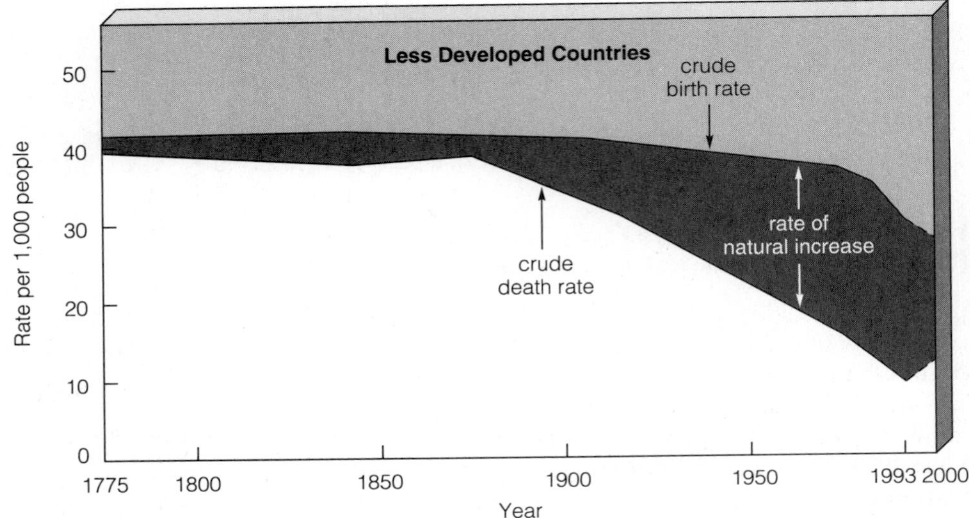

period who are still moving through their childbearing years (even though the total fertility rate has remained at or below replacement level for 23 years, there has been a large increase in the number of potential mothers) and **(2)** high levels of legal and illegal immigration.

CONNECTIONS: FACTORS AFFECTING BIRTH RATES AND FERTILITY RATES The most significant interrelated factors affecting a country's average birth rate and total fertility rate are:

- *Average level of education and affluence.* Rates are usually lower in MDCs, where levels of both education and affluence are higher than in LDCs.
- *Importance of children as a part of the family labor force.* Rates tend to be lower in MDCs and higher in LDCs (especially in rural areas).
- *Urbanization.* People living in urban areas usually have better access to family planning services and tend to have fewer children than those living in rural areas, where children are needed to help grow food, collect firewood and water, and perform other essential tasks.
- *Cost of raising and educating children.* Rates tend to be lower in MDCs, where raising children is much more costly because children don't enter the labor force until they reach their late teens or early twenties.
- *Educational and employment opportunities for women.* Rates tend to be low when women have access to both education and paid employment outside the home.
- *Infant mortality rate.* In areas with low infant mortality rates, people tend to have fewer children

Q: How much of the hazardous waste produced in the United States is regulated by federal laws?

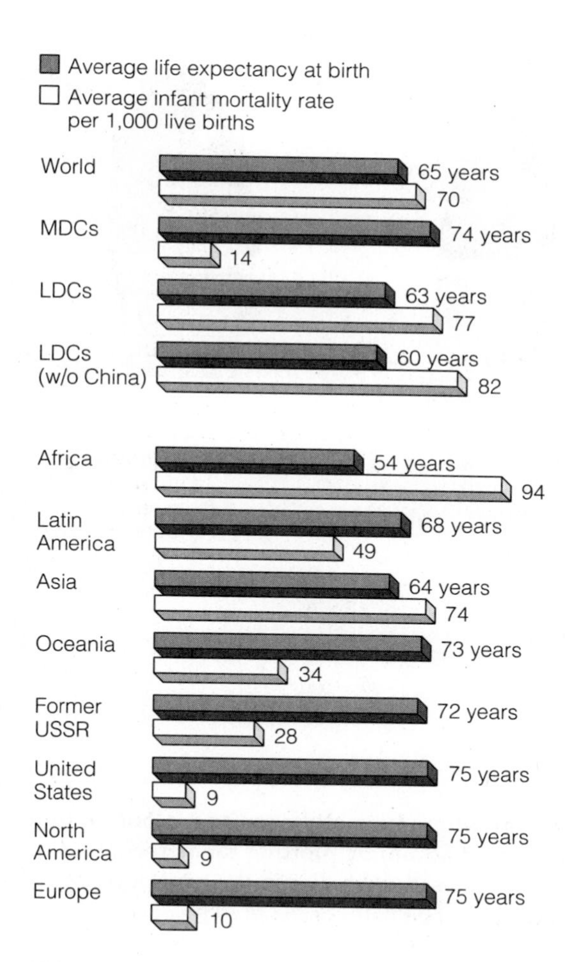

Figure 2-4 Average life expectancy at birth and average infant mortality rate for various groups of countries in 1993. (Data from Population Reference Bureau)

because they don't need to replace children who have died.

- *Average age at marriage* (or more precisely, the average age at which women give birth to their first child). People have fewer children when women's average age at marriage is 25 or older.
- *Availability of private and public pension systems.* Pensions eliminate the need for parents to have many children to support them in old age.
- *Availability of reliable methods of birth control.* Widespread availability tends to reduce birth and fertility rates.
- *Religious beliefs, traditions, and cultural norms that influence the number of children couples want to have.* In many LDCs these factors favor large families.

CONNECTIONS: FACTORS AFFECTING DEATH RATES The rapid growth of the world's population over the past 100 years was not caused by a rise in crude birth rates. Rather, it is due largely to a decline in crude death rates, especially in the LDCs (Figure 2-3). The principal interrelated reasons for this general drop in death rates are:

- *Better nutrition* because of greater food production and better food distribution
- *Fewer infant deaths and longer average life expectancy* due to improved personal hygiene, sanitation, and water supplies, which have curtailed the spread of many infectious diseases
- *Improvements in medical and public health technology*, including antibiotics, immunizations, and insecticides

Two useful indicators of overall health in a country or region are **life expectancy**—the average number of years a newborn infant can be expected to live—and the **infant mortality rate**—the number of babies out of every 1,000 born each year that die before their first birthday (Figure 2-4). In most cases a low life expectancy in a particular area is the result of high infant mortality.

Because it reflects the general level of nutrition and health care, infant mortality is probably the single most important measure of a society's quality of life. A high infant mortality rate usually indicates insufficient food (undernutrition), poor nutrition (malnutrition), and a high incidence of infectious disease (usually from contaminated drinking water). Between 1965 and 1993, the world's infant mortality rate dropped 31% in MDCs and 35% in LDCs. This is an impressive achievement, but it still means that at least 12 million infants die each year of preventable causes. Although the U.S. infant mortality rate of 8.6 per 1,000 in 1993 was low by world standards, 23 other countries had lower rates. Several factors keep the U.S. infant mortality rate higher than it could be; they include:

- Inadequate health care for poor women during pregnancy and for their babies after birth.
- Drug addiction among pregnant women.
- The high birth rate among teenage women. The United States has the highest teenage pregnancy rate of any MDC. Among U.S. teenage women there are about 460,000 abortions and 490,000 births per year. Babies born to teenagers are more likely to have a low birth weight, the most important factor in infant deaths.

MIGRATION The population change for a specific geographic area is also affected by movement of people into (immigration) and out of (emigration) that area, according to the following equation:

A: 6%

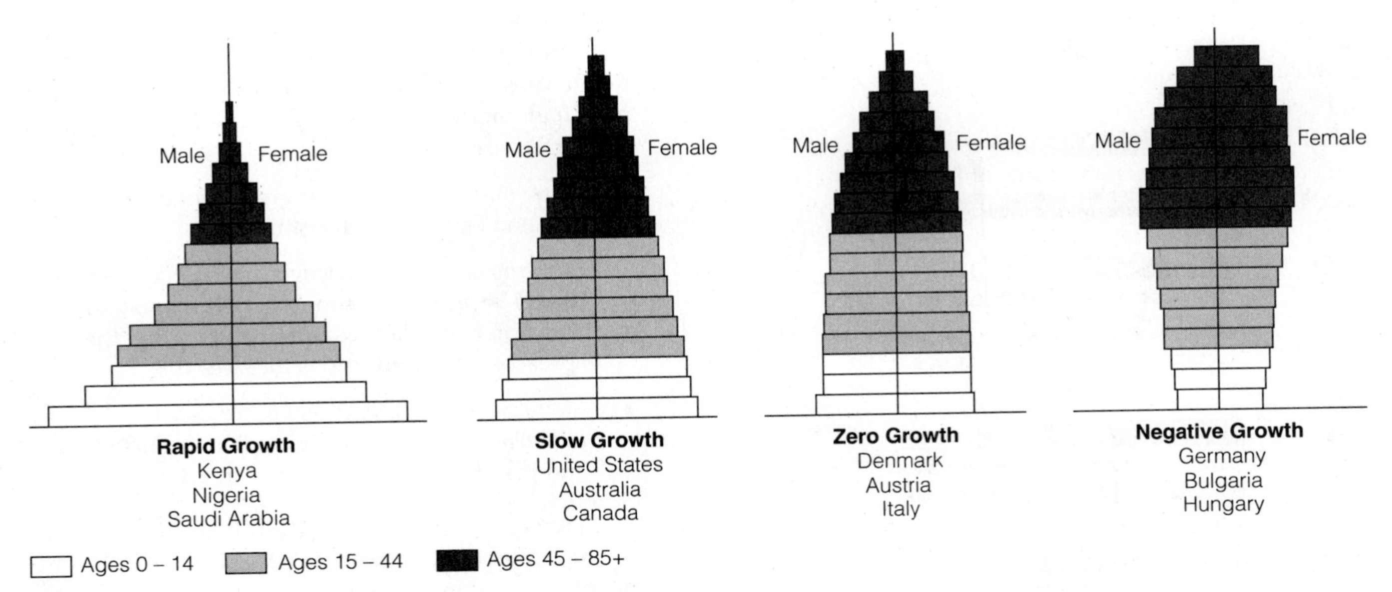

Figure 2-5 Population age structure diagrams for some countries with rapid, slow, zero, and negative population growth rates. Bottom portions of the diagrams represent prereproductive years (ages 0–14), middle portions represent reproductive years (ages 15–44), and top portions represent postreproductive years (ages 45–85+). (Data from Population Reference Bureau)

$$\text{Population Change Rate} = \left(\begin{array}{c}\text{Births} \\ + \\ \text{Immigration}\end{array}\right) - \left(\begin{array}{c}\text{Deaths} \\ + \\ \text{Emigration}\end{array}\right)$$

Most countries control their rates of population growth to some extent by restricting immigration; only a few countries accept large numbers of immigrants or refugees. Thus population change for most countries is determined mainly by the difference between their birth and death rates. Migration within countries, especially from rural to urban areas, plays an important role in the population dynamics of cities, towns, and rural areas, as discussed in Section 2-3.

2-2 Population Momentum: Age Structure

AGE STRUCTURE DIAGRAMS Even if the replacement-level fertility rate of 2.1 were magically achieved globally by tomorrow, the world's population would keep on growing for at least another 60 years! Why? The answer lies in an understanding of the **age structure**, or age distribution, of a population—that is, the percentage of the population (or the number of people of each sex) at each age level.

Demographers typically construct a population age structure diagram by plotting the percentages or numbers of males and females in the total population in each of three age categories: *prereproductive* (ages 0–14), *reproductive* (ages 15–44), and *postreproductive* (ages 45–85+). Figure 2-5 shows the age structure diagrams for countries with rapid, slow, zero, and negative population growth rates.

CONNECTIONS: AGE STRUCTURE AND POPULATION GROWTH MOMENTUM Any country with many people below age 15 (represented by a wide base in Figure 2-5) has powerful built-in momentum toward an increase in population size unless death rates rise sharply. The number of births rises even if women have only one or two children because of the large number of women moving into their reproductive years.

Today half of the world's 2.7 billion women are in the reproductive age category (15–44), and one-third of the people on this planet (36% in LDCs and 21% in MDCs) are under 15 years old and are poised to move into their prime reproductive years. Even if each female in LDCs has only two children, world population will still grow for 60 years unless the death rate rises sharply. And women in LDCs now produce an average of 3.7 children, well above replacement level. This powerful force for continued population growth, mostly in LDCs, will be slowed only by an effective program to reduce birth rates or by a catastrophic rise in death rates.

Q: What two countries have the world's largest populations?

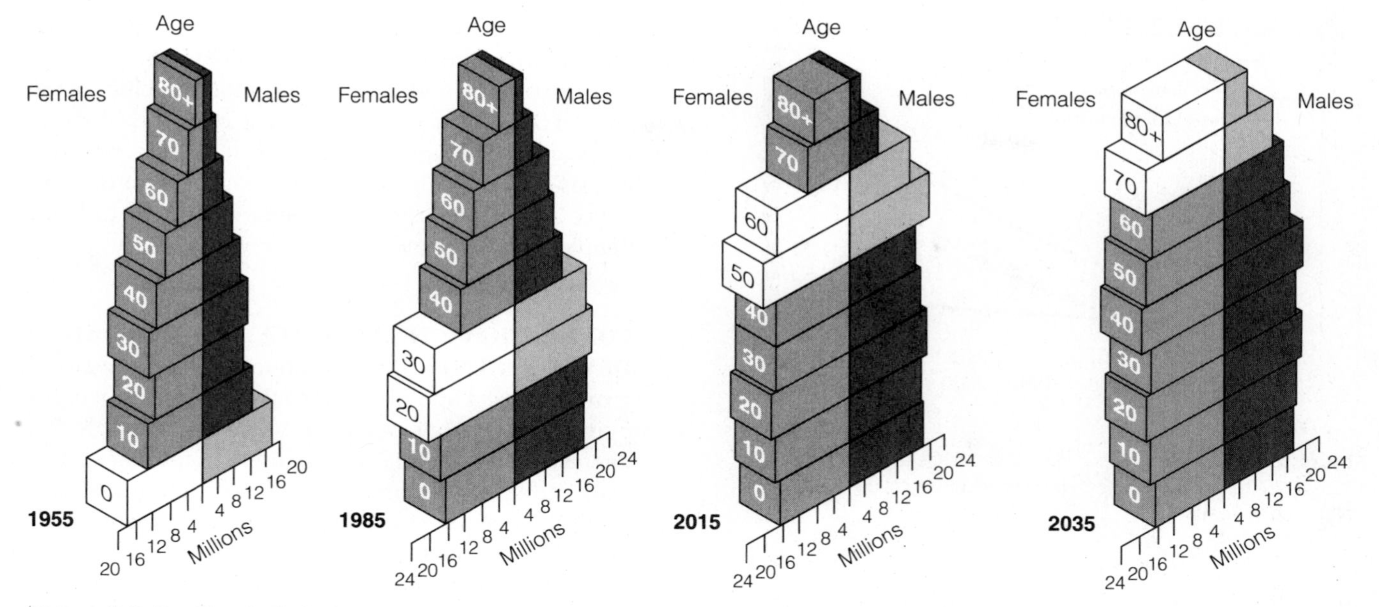

Figure 2-6 Tracking the baby-boom generation in the United States. (Data from Population Reference Bureau and U.S. Census Bureau)

CONNECTIONS: MAKING PROJECTIONS FROM AGE STRUCTURE DIAGRAMS A large increase occurred in the U.S. population between 1946 and 1964 (Figure 2-2). This 80-million-person bulge, known as the *baby-boom generation,* will move upward through the country's age structure between 1946 and 2040 as baby boomers leave young adulthood and enter their middle and then old age (Figure 2-6).

Today baby boomers make up nearly half of all adult Americans. They dominate the population's demand for goods and services and play an increasingly important role in deciding who gets elected and what laws are passed.

The economic burden of helping support so many retired baby boomers will fall on the *baby-bust generation,* the much smaller group of 47 million people born in the period between 1970 and 1985, during which total fertility rates fell sharply (Figure 2-2). Retired baby boomers may use their political clout to force members of the baby-bust generation to pay higher income, health care, and Social Security taxes.

In other respects the baby-bust generation should have an easier time than the baby-boom generation. Fewer people will be competing for educational opportunities, jobs, and services, and labor shortages may drive up their wages, at least for jobs requiring education or technical training beyond high school. On the other hand, members of the baby-bust group may find it difficult to get job promotions as they reach middle age because most upper-level positions will be occupied by members of the much larger baby-boom group. And many baby boomers may delay retirement because of improved health and the need to accumulate adequate retirement funds.

From these few projections we can see that any booms or busts in the age structure of a population create social and economic changes that ripple through a society for decades.

2-3 Population Distribution: Urbanization and Urban Problems

THE FUTURE IS URBAN Since 1950 the number of people living in urban areas has more than tripled and by 2025 is projected to reach 5.5 billion—equal to the world's current population (Figure 2-7). Today about 42% of the world's population lives in urban areas, and by 2025 this figure is expected to increase to 61%. During the 1990s about 83% of the world's population increase is expected to occur in urban areas. At current rates the world's population will double in 41 years, the urban population in 22 years, and the urban population of LDCs in only 15 years.

People are drawn to urban areas in search of jobs and a better life. They may also be pushed into urban areas by modern mechanized agriculture, which uses fewer farm laborers and allows large landowners to buy out subsistence farmers who cannot afford to

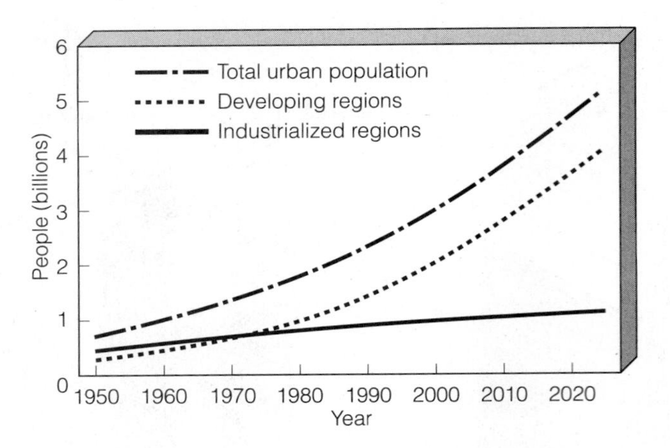

Figure 2-7 Urban population growth in MDCs and LDCs, 1950–2025. (Data from United Nations and Population Reference Bureau)

modernize. Without jobs or land, these people are forced to move to cities. The poor people fortunate enough to get jobs must usually work long hours for low wages. These jobs may also expose them to dust, hazardous chemicals, excessive noise, and dangerous machinery.

The number of large cities is mushrooming. Today 1 person out of every 10 lives in a city with a million or more inhabitants, and many live in *megacities* with 10 million or more people. The United Nations projects that by 2000 there will be 26 megacities, more than two-thirds of them in LDCs.

Poverty is becoming urbanized as more poor people migrate from rural to urban areas. At least 1 billion people—18% of the world's population—live in the crowded slums of inner cities and in the vast, mostly illegal squatter settlements or shantytowns that ring the outskirts of most cities in LDCs. Despite being centers of commerce and industry, many cities suffer from extreme poverty and social and environmental decay.

POPULATION DISTRIBUTION IN THE UNITED STATES In 1800 only 5% of Americans lived in cities; now about 75% of Americans live in the nation's 350 *metropolitan areas*—cities and towns with at least 50,000 people. Nearly half of the country's population lives in large metropolitan areas containing 1 million or more residents (Figure 2-8).

Since 1920, many of the worst urban environmental problems in the United States (and other MDCs) have been significantly reduced. Most people have better working and housing conditions; air and water quality have improved. Better sanitation, public water supplies, and medical care have slashed death rates and the prevalence of sickness from malnutrition and transmittable diseases such as measles, diphtheria, typhoid fever, pneumonia, and tuberculosis.

The biggest problems facing numerous cities in the United States and other MDCs are deteriorating services, aging infrastructures (streets, schools, bridges, housing, sewers), budget crunches from lost tax revenues and rising costs as businesses and more affluent people move out, and increased crime.

CONNECTIONS: TRANSPORTATION AND URBAN DEVELOPMENT If a city cannot spread outward, it grows upward and downward (below ground), occupying a relatively small area with a high population density. Most people living in such compact cities walk, ride bicycles, or use energy-efficient mass transit. Residents often live in multistory apartment buildings; with few outside walls in many apartments, heating and cooling costs are reduced. Many European cities and urban areas such as Hong Kong and Tokyo are compact and tend to be more energy-efficient than the dispersed cities in the United States, Canada, and Australia, where ample land for outward expansion often exists.

A combination of cheap gasoline, plentiful land, and a network of highways leads to dispersed, car-oriented cities with a low population density—a condition often called *urban sprawl*. Most people living in such urban areas live in single-family houses, with unshared walls that lose and gain heat rapidly unless they are well insulated and airtight. Urban sprawl also gobbles up unspoiled natural habitats and paves over fertile farmland, and it promotes considerable reliance on automobiles.

The United States, with only 4.7% of the world's people, has 35% of the world's cars and trucks. In the United States cars are used for 86% of all trips (compared to about 45% in most western-European countries), for 98% of all urban transportation, and for 86% of travel to work. Almost 75% of commuting cars carry only one person, and only 13% of commuters use carpools. Only about 5% of Americans use public transportation, and only 7% walk or use a bicycle to get to and from work. Each year Americans drive as far as the rest of the world combined. No wonder the British author J. B. Priestley remarked, "In America, the cars have become the people."

The automobile provides convenience and undreamed-of mobility. To many people cars are also symbols of power, sex, excitement, and success. Moreover, much of the world's economy is built on producing motor vehicles and supplying roads, services, and repairs for them. In the United States, one of every six dollars spent and one of every six nonfarm jobs are connected to the automobile.

Q: How much of the world's population is in the United States?

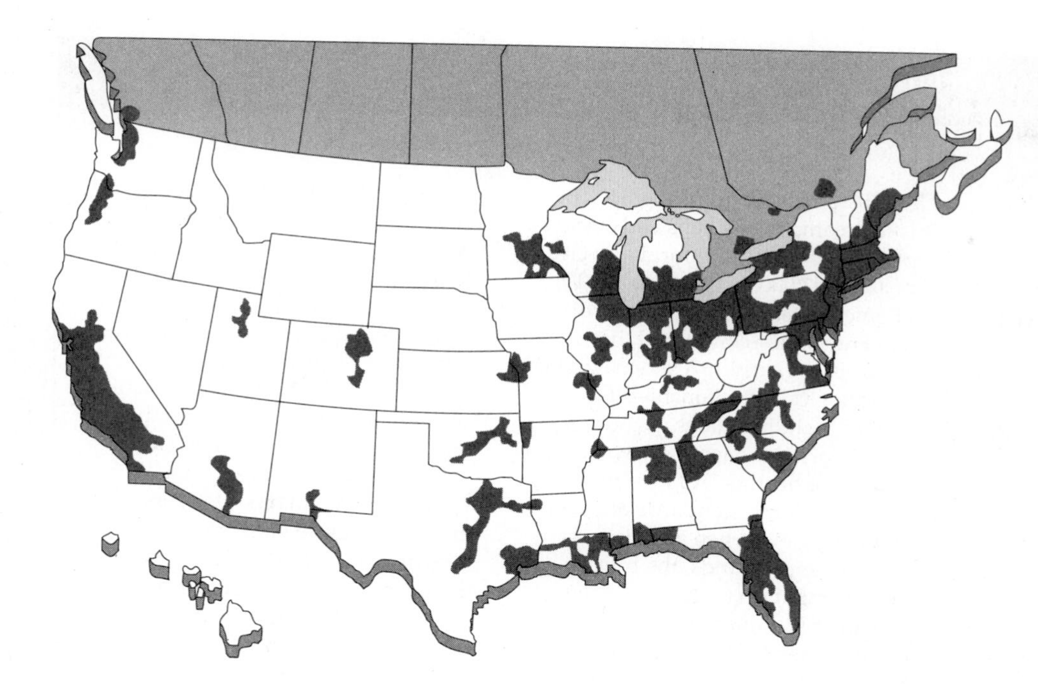

Figure 2-8 Major urban regions in the United States by 2000. Nearly half (48%) of Americans live in *consolidated metropolitan areas* containing 1 million or more people. (Data from U.S. Census Bureau)

In spite of their important economic and personal benefits, motor vehicles have many destructive effects on human lives and on air, water, land, and wildlife resources. Since 1885, when Karl Benz built the first automobile, almost 18 million people have been killed by motor vehicles. This death toll increases by about 250,000 people per year (40,000 in the United States), and each year about 10 million people are injured or permanently disabled in motor vehicle accidents. More Americans have been killed by cars than were killed in all the country's wars.

Motor vehicles are also the largest source of air pollution, laying a haze of smog over the world's car-clogged cities. In the United States they produce at least 50% of the air pollution, even though emission standards are as strict as any in the world.

By making long commutes and distant shopping possible, automobiles and highways have helped create urban sprawl and reduced use of mass transit, bicycling, and walking. Worldwide, at least a third of urban land is devoted to roads, parking lots, gasoline stations, and other automobile-related uses. Urban expert Lewis Mumford has suggested that the U.S. national flower should be the concrete cloverleaf.

In 1907, the average speed of horse-drawn vehicles through the borough of Manhattan was 18.5 kilometers (11.5 miles) per hour. Today cars and trucks with the potential power of 100–300 horses creep along Manhattan streets at 8 kilometers (5 miles) per hour. In Paris and Tokyo, average auto speeds are even lower. If current trends continue, U.S. motorists will spend an average of two years of their lifetimes in traffic jams. Building more roads is not the answer because, as economist Robert Samuelson put it, "cars expand to fill available concrete."

One way to break this cycle is full-cost pricing for automobiles (Solutions, p. 5). Federal, state, and local government auto subsidies in the United States amount to at least $300 billion a year—an average of $1,600 per vehicle. If drivers had to pay these hidden costs directly as a gasoline tax, the tax on each gallon would be about $5. Taxpayers (drivers and nondrivers alike) are paying these hidden costs anyway but don't associate them with driving. Although such full-cost pricing makes economic and environmental sense, this approach faces massive political opposition from both the general public and the powerful automobile and petroleum industries.

MAKING CITIES MORE SUSTAINABLE Most of today's cities aren't sustainable, but rather are heavily dependent on distant sources for their food, water, energy, and materials. Their massive use of resources damages nearby and distant air, water, soil, and wildlife.

Sharply decreasing the urban populations and spreading people out over the countryside is not a solution. Even if economically feasible, this would put massive pressure on already-stressed soils, forests, grasslands, wetlands, and other forms of Earth capital. Thus, an important goal in coming decades should be to make urban areas more sustainable and enjoyable places to live (Solutions, p. 28).

A: 4.7% (258 million people in mid-1993)

Is Your City Green?—The Ecocity Concept in Davis, California

SOLUTIONS

In a sustainable and ecologically healthy city—called an *ecocity* or a *green city*—emphasis is placed on efficient use of energy and matter resources, pollution prevention, and reuse, recycling, and composting of 60–80% of the solid waste produced. Trees and plants adapted to the local climate and soils are planted throughout an ecocity to provide shade and beauty, to reduce pollution and noise, and to supply habitats for wildlife. Abandoned lots and polluted creeks are cleaned up and restored. Nearby forests, grasslands, wetlands, and farms are preserved instead of being devoured by urban sprawl. Much of an ecocity's food comes from nearby organic farms, solar greenhouses, community gardens, and small gardens on rooftops and in yards and window boxes. An ecocity is a people-oriented city—not a car-oriented city. Its residents are able to walk or bike to most places, including work, and to take low-polluting mass transit.

The ecocity concept is not a futuristic dream. The citizens and elected officials of Davis, California—a city of about 47,000 people about 130 kilometers (80 miles) northeast of San Francisco—committed themselves in the early 1970s to making it an ecologically sustainable city.

City building codes encourage the use of solar energy for water and space heating. All new homes must meet high standards of energy efficiency, and when an existing home changes hands, the buyer must bring it up to the energy conservation standards for new homes. Development of the surrounding fertile farmland for residential or commercial use is restricted. In Davis's Village Homes—America's first solar neighborhood—houses are heated by solar energy. They face into a common open space reserved for people and bicycles; cars are restricted to streets, which are located only on the periphery of the development. The neighborhood also has orchards, vineyards, and a large community garden.

Since 1975 the city has cut its use of energy for heating and cooling in half. Davis has a solar power plant; some of the electricity it produces is sold to the regional utility company. Eventually the city plans to generate all of its own electricity.

The city discourages the use of automobiles and encourages the use of bicycles by closing some streets to automobiles, by including bike lanes on major streets, and by building bicycle paths. Any new housing tract must have a separate bike lane, and some city employees are given bikes. As a result, 28,000 bicycles account for 40% of all in-city transportation, and less land is needed for parking spaces. This heavy dependence on the bicycle is aided by the city's warm climate and flat terrain. What is the place where you live doing to make life more sustainable and enjoyable?

Solutions: Influencing Population Size

A society can influence the size and rate of growth or decline of its population by encouraging a change in any of the three basic demographic variables: migration, births, and deaths. In the following sections we consider the first two of these variables.

CONTROLLING MIGRATION Only a few countries, chiefly Canada, Australia, and the United States, allow large annual increases in population from immigration, and some countries encourage emigration to reduce population pressures.

Between 1960 and 1993, the annual number of legal immigrants in the United States rose from 265,000 to about 1 million, which is twice as many legal immigrants as all other countries combined. Each year 200,000–500,000 more people enter illegally, most from Mexico and other Latin American countries. Thus in 1993, legal and illegal immigration accounted for 37–43% of the country's population growth. Soon, immigration is expected to be the primary factor increasing the population of the United States.

Some demographers and environmentalists call for an annual ceiling of no more than 450,000 people for all categories of legal immigration, including refugees. They argue that this will allow the United States both to stabilize its population size sooner and reduce the country's enormous environmental impact from consumption overpopulation (Figure 1-12). Others oppose such limits, arguing that it would diminish the historical role of the United States as a place of opportunity for the world's poor and oppressed.

Q: What percentage of the world's population is under age 15?

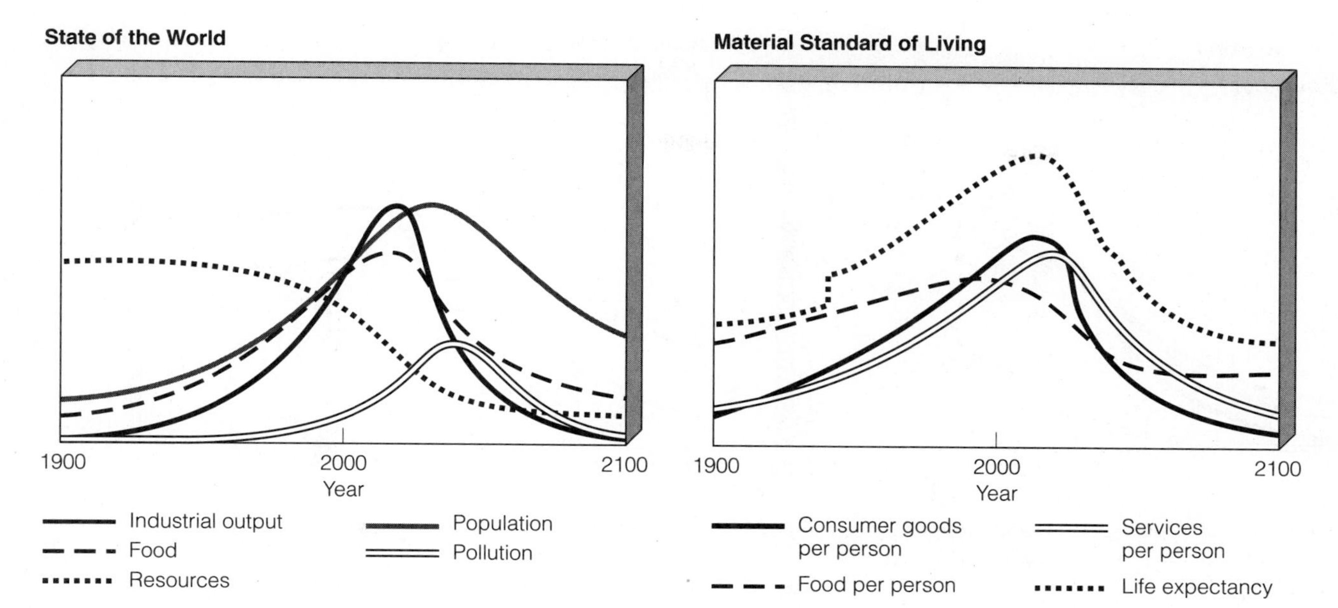

Figure 2-9 Computer model projecting what might happen if the world's population and economy continue growing exponentially at 1990 levels, assuming no major policy changes or technological innovations. This scenario indicates that the world has already overshot some of its limits, and it projects that if current trends continue unchanged, we face global economic and environmental collapse sometime in the twenty-first century. (Used by permission from Donella Meadows et al., *Beyond the Limits: Confronting Global Collapse, Envisioning a Sustainable Future,* Post Mills, Vt.: Chelsea Green Publishing Co., 1992)

REDUCING BIRTHS: A CONTROVERSIAL ISSUE
Because raising the death rate is not ethically acceptable, lowering the birth rate is the focus of most efforts to slow population growth. Today about 93% of the world's population and 91% of the people in LDCs live in countries with fertility reduction programs.

There is intense controversy over whether the world is overpopulated and over measures used to control population growth. Those who disagree with the view that the earth is overpopulated point out that the average life span of the world's 5.5 billion people is longer than at any time in the past; thus things are getting better, not worse, for many of the world's people. Those holding this view say that talk of a population crash is alarmist, that the world can support billions more people, and that people are the world's most valuable resource for solving the problems we face. Having more people increases economic productivity by creating and applying new knowledge, and people stimulate economic growth by becoming consumers.

Some people view population regulation as a violation of their deep religious beliefs, whereas others see it as an intrusion into their privacy and personal freedom. They believe that all people should be free to have as many children as they want. And some members of minorities regard population control as a form of genocide to keep their numbers and power from rising.

Proponents of slowing and eventually stopping population growth point out that currently we are not providing adequate basic necessities for one person out of five on Earth (Spotlight, p. 7). They also view people overpopulation in LDCs and consumption overpopulation in MDCs (Figure 1-12, page 14) as threats to Earth's life-support systems. Although they concede that population growth is not the only cause of our environmental and resource problems, they contend that adding several hundred million more people in MDCs and several billion more in LDCs will intensify many environmental and social problems (Figure 2-9) by increasing resource use and waste, increasing environmental degradation and pollution, increasing the threat of climate change, and reducing biodiversity, all of which could lead to economic and political chaos. They call for sharply cutting population growth and making other drastic changes to prevent accelerating environmental decline, the effects of which are projected in Figure 2-10.

These analysts believe that people should have the freedom to produce as many children as they want only so long as this freedom does not reduce the quality of other people's lives now and in the future by

A: 33% (36% in LDCs and 21% in MDCs) in 1993

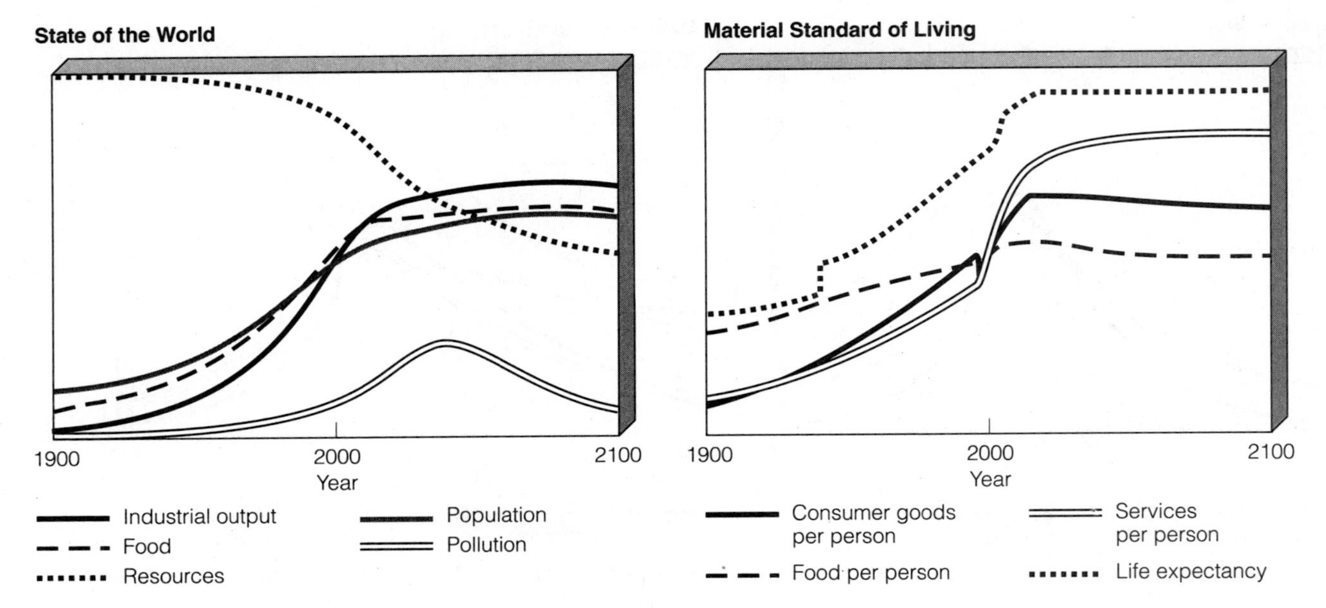

Figure 2-10 Computer model projecting how we can avoid overshoot and collapse and make a fairly smooth transition to a sustainable future. This scenario assumes that **(1)** technology allows us to double supplies of nonrenewable resources, double crop and timber yields, cut soil erosion in half, and double the efficiency of resource use within 20 years; **(2)** 100%-effective birth control is made available to everyone by 1995; **(3)** no couple has more than two children beginning in 1995; and **(4)** industrial output per capita is stabilized at 1990 levels. Another computer run of this model projects that waiting until 2015 to implement these changes would lead to collapse and overshoot sometime around 2075, followed by a transition to sustainability by 2100. (Used by permission from Donella Meadows et al., *Beyond the Limits: Confronting Global Collapse, Envisioning a Sustainable Future,* Post Mills, Vt.: Chelsea Green Publishing Co., 1992)

impairing the ability of the earth to sustain life. They point out that limiting an individual's freedom to do certain things to protect the freedom of all is the basis of most laws in modern societies.

REDUCING BIRTHS THROUGH ECONOMIC DEVELOPMENT Demographers have examined the birth and death rates of western-European countries that industrialized during the nineteenth century, and from these data they developed a hypothesis of population change known as **the demographic transition**: As countries become industrialized, first their death rates and then their birth rates decline (Figure 2-11).

In most LDCs today death rates have fallen much more than birth rates. In other words, these LDCs—mostly in Southeast Asia, Africa, and Latin America—are still in the transitional stage (Figure 2-11), halfway up the economic ladder, with high population growth rates. Some economists believe that LDCs will become economically developed and make the demographic transition over the next few decades. Many population analysts, however, fear that the rapid population growth in many LDCs (Figure 1-5) will outstrip economic growth—as evidenced by the growing wealth gap between MDCs and LDCs (Figure 1-6)—and will overwhelm local life-support systems. As a result, many of these countries could be caught in a *demographic trap.*

REDUCING BIRTHS THROUGH FAMILY PLANNING **Family-planning programs** provide educational and clinical services that help couples choose how many children to have and when to have them. They have been an important factor both in increasing the percentage of married women in LDCs using contraception (from 10% in the 1960s to 54% today) and in the drop in total fertility rates in LDCs (from 6 in 1960 to 3.7 in 1993).

Family planning also saves a society's money by reducing the need for various social services. It also has health benefits. In LDCs, for example, about 1 million women die each year from pregnancy-related causes. It is estimated that half of these deaths could be prevented by effective family-planning and health care programs. Family planning could be provided in LDCs to all couples who want it for about $10 billion a year—the equivalent of less than four days' worth of worldwide military expenditures.

Q: How many of the world's people live in urban areas?

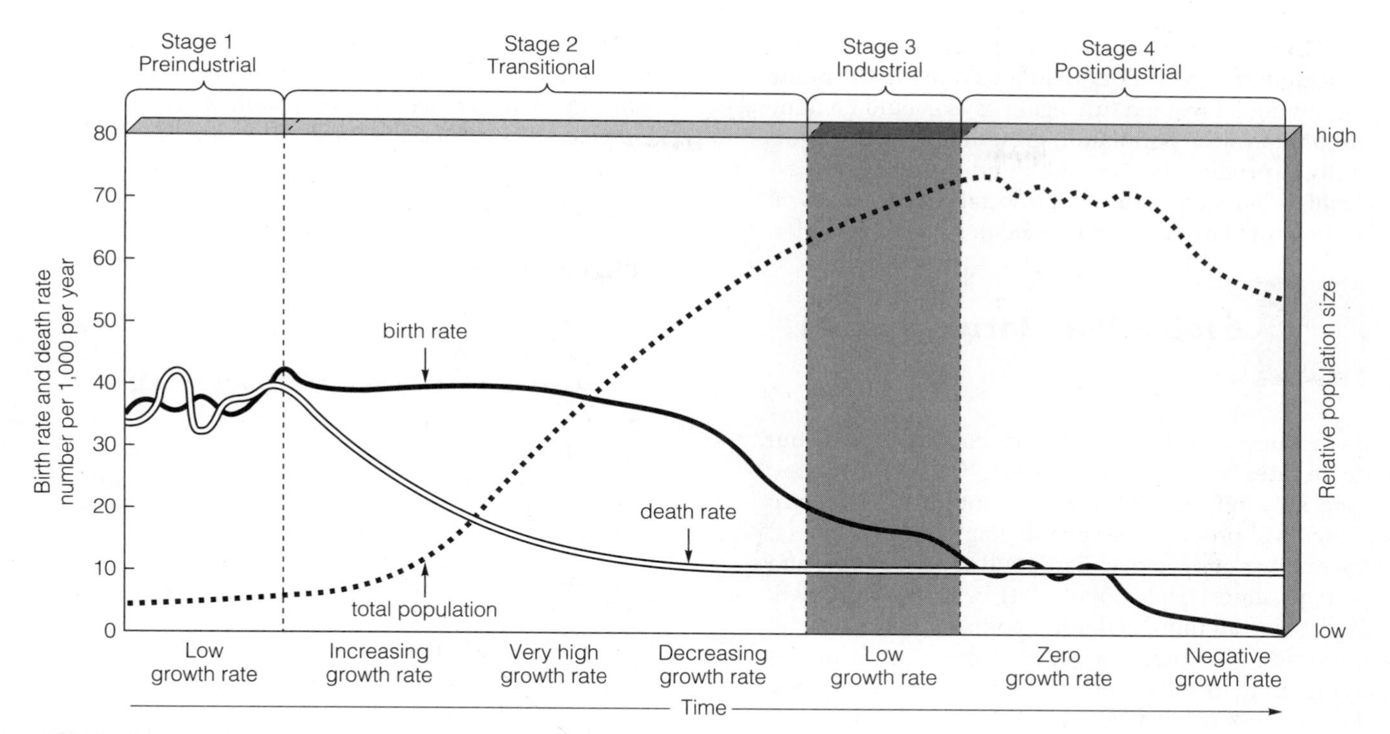

Figure 2-11 Generalized model of the demographic transition.

USING ECONOMIC REWARDS AND PENALTIES TO REDUCE BIRTHS Some population experts argue that family planning, even coupled with economic development, cannot lower birth and fertility rates fast enough to avoid a sharp rise in death rates (Figure 2-9), especially in many LDCs. The main reason for this is that most couples in LDCs want three or four children—well above the 2.1 fertility rate needed to bring about eventual population stabilization.

These analysts believe that we must go beyond family planning and offer economic rewards and penalties to help slow population growth. About 20 countries offer small payments to individuals who agree to use contraceptives or to be sterilized. They also pay doctors and family-planning workers for each sterilization they perform and each IUD they insert. Such payments are most likely to attract people who already have all the children they want, however.

Some countries (for example, China) penalize couples who have more than a certain number of children—usually one or two—by raising their taxes, charging other fees, or not allowing income tax deductions for a couple's third child (as in Singapore, Hong Kong, Ghana, and Malaysia). Families who have more children than the prescribed limit may also lose health care benefits, food allotments, and job choice.

Experience has shown that economic rewards and penalties designed to reduce fertility work best if they nudge rather than push people to have fewer children, reinforce existing customs and trends toward smaller families, do not penalize people who produced large families before the programs were established, and increase a poor family's income or land.

REDUCING BIRTHS BY EXPANDING OPPORTUNITIES FOR WOMEN Research has shown that women tend to have fewer children when they have access to education and to jobs outside the home and when they live in societies where their rights are not suppressed. Today women do almost all of the world's domestic work and child care and provide more health care with little or no pay than all the world's organized health services put together. They also do more than half the work associated with growing food, gathering fuelwood, and hauling water. As one Brazilian woman put it, "For poor women the only holiday is when you are asleep."

Despite their vital economic and social contributions, most women in LDCs don't have a legal right to own land or to borrow money to increase agricultural productivity. Although women work two-thirds of all hours worked in the world, they get only one-tenth of the world's income and own a mere 1% of the world's land. In some LDCs in which male children are strongly favored, young girls are sometimes sold as work slaves or sex slaves or are even killed.

A: 42% (72% in MDCs, 34% in LDCs, 75% in the U.S.) in 1993

Giving more women the opportunity to become educated and to work at paid jobs outside the home and giving all women full legal rights should be done not only to slow population growth but more importantly to promote human rights and freedom. However, this will require some major social changes that will be difficult to bring about in male-dominated societies.

2-5 Global Population Trends

Birth rates and death rates are coming down, but death rates have fallen more sharply than birth rates, especially in LDCs. If this trend continues, one of two things will probably happen during your lifetime: **(1)** the number of people on Earth will at least double and perhaps almost triple (Figure 1-1), or **(2)** the world will experience an unprecedented population crash, with hundreds of millions of people—perhaps billions—dying prematurely (Figure 2-9). Lester Brown, president of the Worldwatch Institute, urges world leaders to adopt the goal of cutting world population growth in half during the 1990s by reducing the average global birth rate from 26 to 18 per 1,000 people. The experience of countries such as Japan, Thailand, and China indicate that such a drop in birth rate is possible.

We need the size of population in which human beings can fulfill their potentialities; in my opinion we are already overpopulated from that point of view; not just in places like India and China and Puerto Rico, but also in the United States and western Europe.

George Wald (Nobel Laureate, Biology)

CRITICAL THINKING

1. Why is it rational for a poor couple in India to have six or seven children? What changes might induce such a couple to think of their behavior as irrational?
2. What conditions, if any, would encourage you to rely less on the automobile? Would you regularly travel to school or work by bicycle or motor scooter, on foot, by mass transit, or in a car (or van) pool? Explain.
3. Are there physical limits to growth on Earth? Explain your answer.
4. **a.** Should the number of *legal* immigrants and refugees allowed into the United States each year be sharply reduced? Explain your answer.
   b. Should *illegal* immigration into the United States be sharply decreased? Explain. If so, how would you go about achieving this?

Q: What percentage of the cars carrying people to and from work in the United States have only one passenger?

3 Biodiversity: Sustaining Soils and Producing Food

Below that thin layer comprising the delicate organism known as the soil is a planet as lifeless as the moon.

G. Y. JACKS AND R. O. WHYTE

3-1 Soil: Types, Erosion, and Contamination

SOIL: THE BASE OF LIFE **Soil** is a complex mixture of inorganic materials (clay, silt, pebbles, and sand), decaying organic matter, water, air, and billions of living organisms. Soil forms when formerly living matter decays, when solid rock weathers and crumbles, and when sediments resulting from erosion are deposited. The living organisms (mostly decomposers) in soils and the plants, animals, and microorganisms supported by soils make up the bulk of the earth's biodiversity.

Unless you are a farmer or a gardener, you probably think of soil as dirt—as something you don't want on your hands, clothes, or carpet. Yet your life and the lives of most other organisms depend on soil, especially topsoil. In addition to food, soil provides us with wood, paper, cotton, and medicines, and it helps purify the water we drink and decompose and recycle biodegradable wastes. And soil nutrients, eroded naturally from land, support food webs in aquatic systems. Yet since the beginnings of agriculture we have abused this vital, potentially renewable resource. Entire civilizations have collapsed because they mismanaged the topsoil that supported their populations.

Mature soils are arranged in a series of zones called **soil horizons**, each with a distinct texture and composition that varies with different types of soils (Figure 3-1). Most mature soils have at least three of the possible horizons. Five important soil types, each with a distinct profile, are shown in Figure 3-2. Most of the world's crops are grown on soils exposed when grasslands and deciduous forests are cleared.

SOIL EROSION **Soil erosion** is the movement of soil components, especially surface-litter and topsoil, from one place to another. The two main agents of erosion are flowing water (by far the major cause of erosion) and wind.

Some soil erosion is natural. In undisturbed vegetated ecosystems, the roots of plants help anchor the soil, and usually soil is not lost faster than it forms. However, farming, logging, building, overgrazing by livestock, off-road vehicles, fire, and other activities that destroy plant cover leave soil vulnerable to erosion.

Losing topsoil makes a soil less fertile and less able to hold water. The resulting sediment, the largest source of water pollution, clogs irrigation ditches, boat channels, reservoirs, and lakes. The water becomes cloudy and tastes bad, fish die, and flood risk increases.

Soil, especially topsoil, is classified as a potentially renewable resource because it is continuously regenerated by natural processes. However, in tropical and temperate areas it takes 200–1,000 years for 2.54 centimeters (1 inch) of new topsoil to form, depending on climate and soil type. If topsoil erodes faster than it forms on a piece of land, the soil there becomes a nonrenewable resource.

Today topsoil is eroding faster than it forms on about one-third of the world's cropland. Each year we must feed 90 million more people with an estimated 24 billion metric tons (26 billion tons) less topsoil. The topsoil that each year washes and blows from one place to another would fill a train of freight cars long enough to encircle the planet 150 times. The situation is worsening as many farmers in LDCs cultivate lands subject to erosion in an effort to feed themselves.

But erosion is not limited to LDCs. Enough topsoil erodes away each day in the United States to fill a line of dump trucks 5,600 kilometers (3,500 miles) long. Topsoil on cultivated land in the United States is eroding about 16 times faster than it can form. Erosion rates are even higher in heavily farmed regions such as the Great Plains, which has lost one-third or more of its topsoil in the 150 years since it was first plowed. Some of this country's most productive agricultural lands, such as those in Iowa, have lost about half their topsoil. California's soil is eroding 80 times faster than it can be formed. Soil expert David Pimentel estimates that the direct and indirect costs of soil erosion and runoff in the United States exceed $25 billion per year—an average loss of $2.9 million per hour!

A: Over 66%

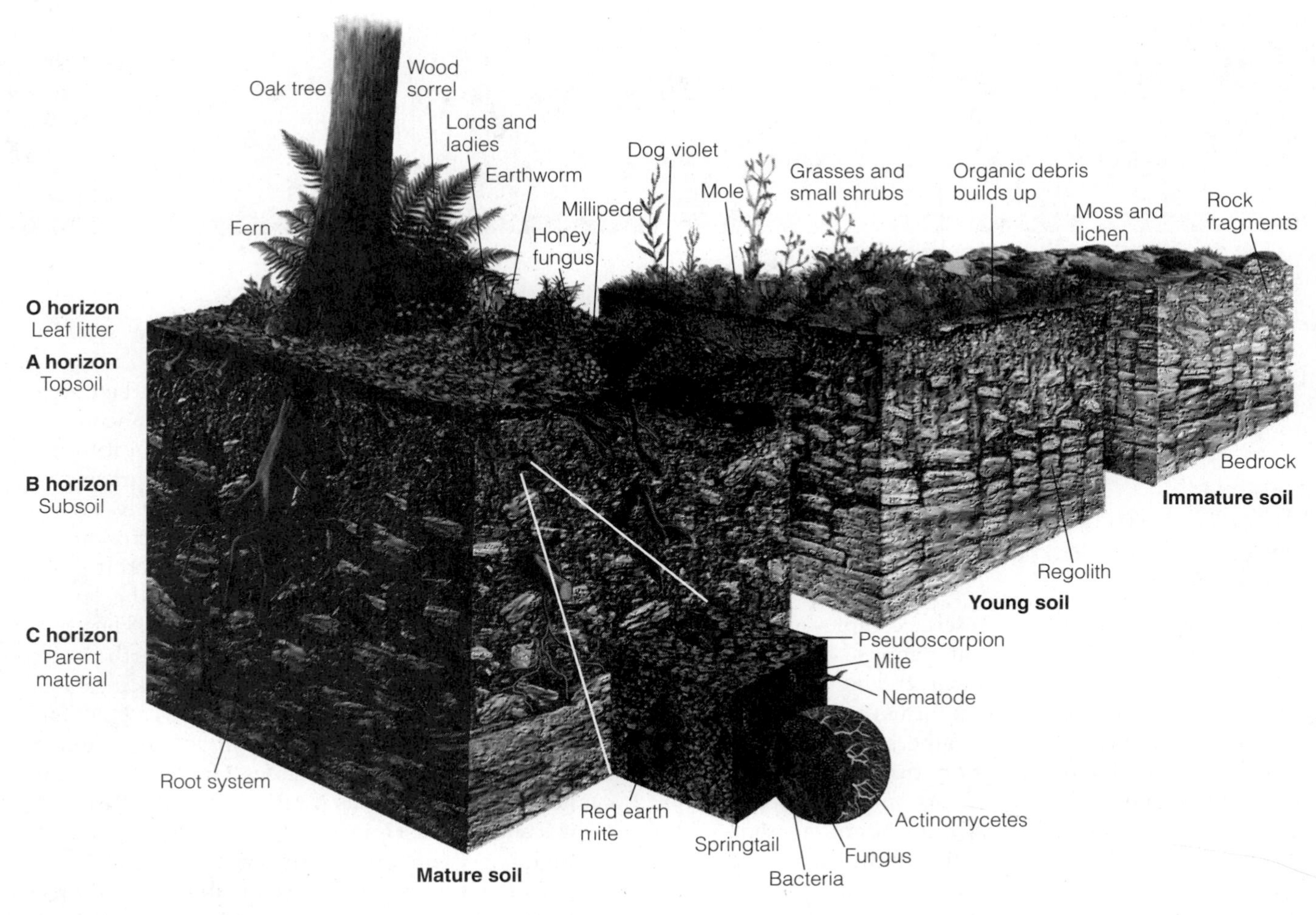

Figure 3-1 Formation and generalized profile of soils. Horizons, or layers, vary in number, composition, and thickness, depending on the type of soil. (Used by permission of Macmillan Publishing Company from Derek Elsom, *Earth*, New York: Macmillan, 1992. Copyright © 1992 by Marshall Editions Developments Limited)

Responsibility for soil erosion is not limited to farmers. At least 40% of soil erosion in the United States is caused by timber cutting, overgrazing, mining, and urban development carried out without proper regard for soil conservation.

SOIL CONTAMINATION BY EXCESS SALTS AND WATER The approximately 18% of the world's cropland that is now irrigated produces about one-third of the world's food. Irrigated land can produce crop yields that are two to three times greater than those from rain-watering, but irrigation has its downside: Irrigation water contains dissolved salts. In dry climates, much of the water in this saline solution evaporates, leaving behind its salts, such as sodium chloride, in the topsoil. The accumulation of these salts, called *salinization*, stunts crop growth, lowers yields, and eventually kills crop plants and ruins the land.

It is estimated that salinization is reducing yields on one-fourth of the world's irrigated cropland. Worldwide, 50–65% of all currently irrigated cropland will probably have undergone reduced productivity from salinization by 2000.

Salts can be flushed out of soil by applying much more irrigation water than is needed for crop growth, but this practice increases pumping and crop-production costs, and it wastes enormous amounts of water. Heavily salinized soil can also be renewed by taking the land out of production for two to five years, installing an underground network of perforated drainage pipes, and flushing the soil with large quantities of low-salt water. This costly scheme, however, only slows the salt buildup; it does not stop the process.

Another problem with irrigation in some areas is *waterlogging*. Farmers often apply heavy amounts of irrigation water to leach salts deeper into the soil.

Q: What percentage of the people use public transportation to get to and from work in the United States?

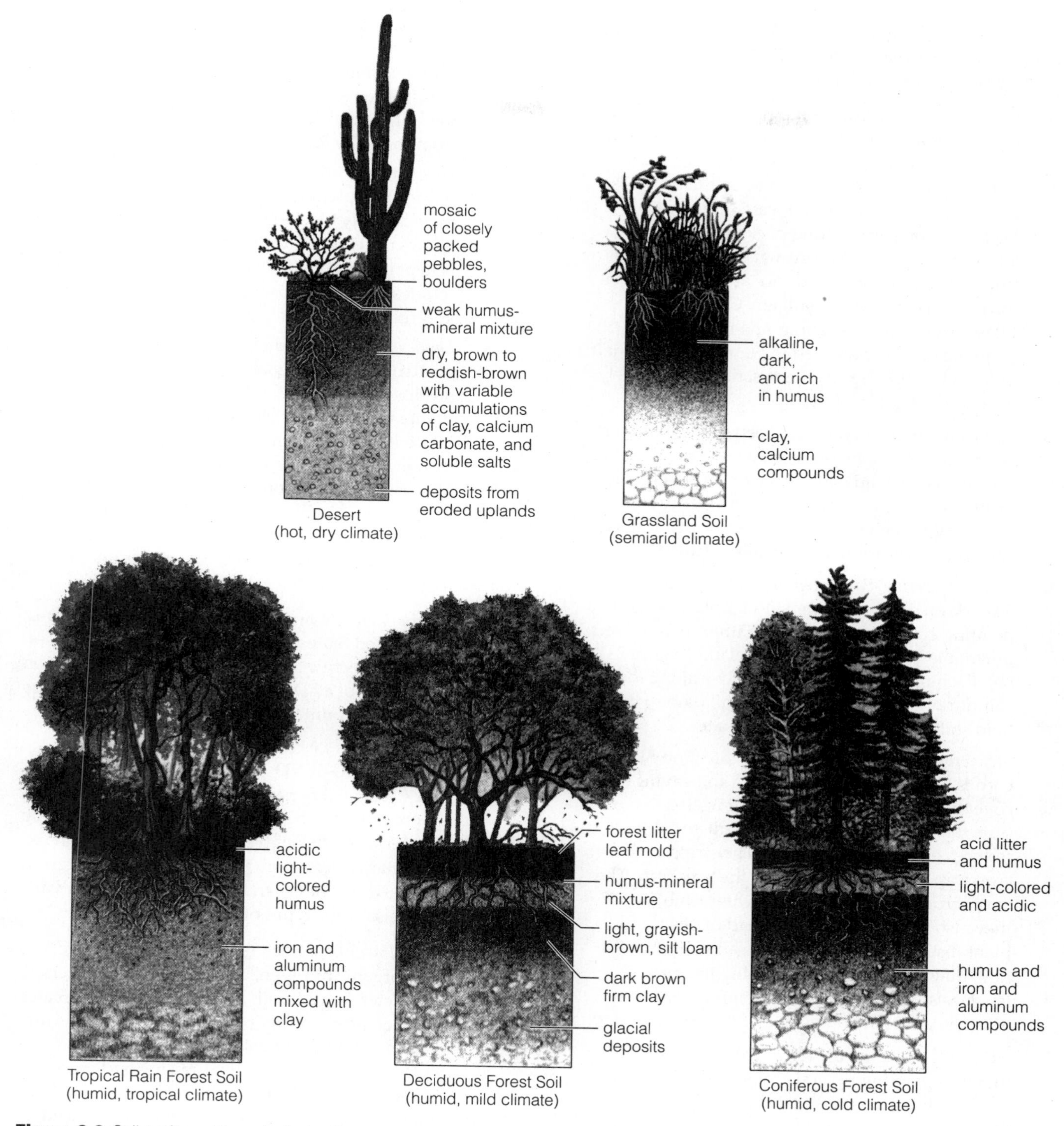

Figure 3-2 Soil profiles of the principal soil types typically found in five different types of biomes.

Without adequate drainage, however, water accumulates underground, gradually raising the water table. Saline water then envelops the roots of plants and kills them. At least one-tenth of all irrigated land worldwide suffers from waterlogging, and the problem is getting worse.

3-2 Solutions: Soil Conservation

REDUCING EROSION **Soil conservation** involves reducing soil erosion and maintaining and restoring

A: 5% (compared to 47% in Japan)

soil fertility. Most methods used to control soil erosion involve keeping the topsoil covered with vegetation. Ways to reduce erosion include:

- *Conservation-tillage farming*. Special tilling and planting machines are used to disturb the soil as little as possible while planting crops. Besides reducing soil erosion, conservation tillage saves fuel, cuts costs, holds more water in the soil, keeps the soil from getting packed down, and allows more crops to be grown during a season (multiple cropping). Yields are at least as high as those from conventional tillage. Conservation tillage is now used on about one-third of U.S. croplands and is projected to be used on over half of them by 2000. So far, the practice is not widely used in other parts of the world.
- *Terracing*. In this practice, a steep slope is converted into a series of broad, nearly level terraces that run across the land contour, with short vertical drops from one terrace to another (Figure 3-3a). Terracing retains water for crops on each terrace and cuts soil erosion by controlling runoff.
- *Contour farming*. Soil erosion can be reduced 30–50% on gently sloping land by plowing and planting crops in rows across, rather than up and down, the sloped contour of the land (Figure 3-3b). Each row planted horizontally—along the contour of the land—acts as a small dam to help hold soil and slow the runoff of water.
- *Strip cropping*. In this practice, a row crop like corn is alternated in strips with a soil-saving cover crop, such as a grass or a grass-legume mixture, that completely covers the soil and thus reduces erosion (Figure 3-3c). The strips of cover crop trap soil that erodes from the row crop. The cover crops catch and reduce water runoff, and they also help prevent the spread of pests and plant diseases from one strip to another. In addition, they help restore soil fertility if legumes (such as soybeans or alfalfa) that replenish soil nitrogen are planted in some of the strips.
- *Alley cropping or agroforestry*. This practice involves planting several crops together in strips or alleys between trees and shrubs that can provide fruit and fuelwood (Figure 3-3d). The trees provide shade (which reduces water loss by evaporation) and help to retain soil moisture and release it slowly. The tree and shrub trimmings can be used as mulch (green manure) for the crops and as fodder for livestock.
- *Gully reclamation*. Small gullies carved out of the land by erosion can be seeded with quick-growing plants such as oats, barley, and wheat for the first season, whereas deeper gullies can be dammed to collect silt and gradually fill in the channels (Figure 3-3e). Fast-growing shrubs, vines, and trees can also be planted to stabilize the soil, and channels can be built to divert water from the gully and prevent further erosion.
- *Windbreaks or shelterbelts*. Wind erosion can be reduced by planting long rows of trees so that they partially block the wind (Figure 3-3f). Windbreaks are especially effective if uncultivated land is kept covered with vegetation; they also help retain soil moisture, supply some wood for fuel, and provide habitats for wildlife.
- *Land-use classification and control*. Land can be evaluated with the goal of identifying easily erodible (marginal) land that should neither be planted in crops nor cleared of vegetation. In the United States, the Soil Conservation Service (SCS) has set up a land-use classification system. The SCS basically relies on voluntary compliance with its guidelines in the almost 3,000 local and state soil and water conservation districts it has established, and it provides technical and economic assistance through local district offices.

Of the world's major food-producing countries, only the United States is reducing some of its soil losses. Even so, effective soil conservation is practiced on only about half of all U.S. agricultural land, and on less than half of the country's most erodible cropland.

MAINTAINING AND RESTORING SOIL FERTILITY Fertilizers partially restore plant nutrients lost by erosion, by crop harvesting, and by leaching when water flows through soil layers. Farmers can use either *organic fertilizer* from plant and animal materials or *commercial inorganic fertilizer* produced from various minerals. Three types of organic fertilizer are:

- *Animal manure*, which improves soil structure, adds organic nitrogen, and stimulates beneficial soil bacteria and fungi. Its use in the United States has decreased, mostly because separate farms for growing crops and raising livestock have replaced most mixed animal-raising and crop-farming operations. In addition, tractors and other motorized farm machinery have replaced horses and other draft animals that naturally added manure to the soil.
- *Green manure*: fresh or growing green vegetation plowed into the soil to increase the organic matter available to the next crop.
- *Compost*: a rich natural fertilizer and soil conditioner that aerates soil, improves its ability to retain water and nutrients, helps prevent erosion,

Q: What percentage of Americans walk or use a bicycle to get to and from work?

a. Terracing

b. Contour Farming

c. Strip Cropping

d. Alley Cropping or Agroforestry

e. Gully Reclamation

f. Windbreaks or Shelterbelts

Figure 3-3 Soil conservation methods.

and prevents nutrients from being wasted in landfills. Farmers and homeowners produce it by piling up alternating layers of carbohydrate-rich plant wastes (such as grass clippings, leaves, weeds, hay, straw, and sawdust), kitchen scraps (such as vegetable remains and egg shells), animal manure, and topsoil.

Today, especially in the United States and other MDCs, farmers rely on commercial inorganic fertilizers to replace or build up soil nutrients. Inorganic

A: 2%

commercial fertilizers are easily transported, stored, and applied. Today, the additional food they help produce feeds one of every three people in the world. Without them, world food output would plummet an estimated 40%.

Commercial inorganic fertilizers have some disadvantages, however. Unless animal manure and green manure are also added, the soil's content of organic matter—and thus its ability to hold water—will decrease, and the soil will become compacted and less suitable for crop growth. In addition, most commercial fertilizers supply only two or three of the 20-odd nutrients needed by plants.

The widespread use of commercial inorganic fertilizers, especially on sloped land near streams and lakes, also causes water pollution as some of the nutrients in the fertilizers are washed into nearby bodies of water. The resulting excess of plant nutrients can cause explosive growth of algae and various plants that use up oxygen dissolved in the water, thereby killing fish. Rainwater seeping through the soil can also leach nitrates in commercial fertilizers into groundwater. Drinking water drawn from wells containing high levels of nitrate can be toxic, especially for infants.

Another method for conserving soil nutrients is *crop rotation*. One year farmers plant areas or strips with nutrient-depleting crops such as corn, tobacco, and cotton; the next year they plant the same areas with legumes, whose root nodules add nitrogen to the soil, or with crops such as oats, barley, rye, or sorghum. This method helps restore soil nutrients and reduces erosion by keeping the soil covered with vegetation.

3-3 How Food Is Produced

Global food production has increased substantially over the past two decades; however, to feed the 8.4 billion people projected by 2025, we must produce and distribute as much food during the next 30 years as was produced in all the years since agriculture began about 10,000 years ago.

The species of plants and animals (species diversity) and the varieties of plants and animals (genetic diversity) used to provide us with food are an important part of the planet's biodiversity. Biologists estimate that the earth has perhaps 30,000 plant species with parts that people can eat. However, just 15 plant and eight animal species supply 90% of our food. Four crops—wheat, rice, corn, and potato—make up more of the world's total food production than all other crops combined.

Grains provide about half the world's calories, with two out of three people eating mainly a vegetarian diet—mostly because they can't afford meat. As incomes rise, people consume even more grain, but indirectly—in the form of meat, eggs, milk, cheese, and other products of domesticated livestock. Although only about one-third of the world's people can afford to eat meat, more than half of the world's cropland (and almost two-thirds of the cropland in the United States) is used to produce livestock feed to supply these individuals with meat. In addition, one-third of the world's fish catch is converted into fish meal to feed livestock consumed by meat eaters in MDCs.

TYPES OF FOOD PRODUCTION There are two major types of agricultural systems: industrialized and traditional. **Industrialized agriculture** uses large amounts of fossil-fuel energy, water, commercial fertilizers, and pesticides to produce huge quantities of one crop or animal for sale. Practiced on about 25% of all cropland, mostly in MDCs, industrialized agriculture has spread since the mid-1960s to some LDCs.

Traditional agriculture consists of two main types. **Traditional subsistence agriculture** typically produces only enough crops or livestock for a farm family's survival; in good years there may be a surplus to sell or to put aside for hard times. Subsistence farmers use human labor and in some cases draft animals. An example of this type of agriculture is a combination of slash-and-burn and shifting cultivation in tropical forests (Figure 3-4). In *slash-and-burn cultivation*, small patches of forest are cleared and the underbrush burned; the ashes help fertilize the nutrient-poor soils found in most tropical forests (Figure 3-2). In *shifting cultivation*, a plot used for several years becomes depleted of nutrients or is reinvaded by the forest. The slash-and-burn growers then move on to and clear a new plot. Each abandoned patch must be left fallow (unplanted) for 10–30 years before the soil becomes fertile enough to grow crops again. This type of cultivation is sustainable only when carried out by relatively small numbers of people widely dispersed in tropical forests. Otherwise, too much of the forest is cleared, and depleted plots are not given enough time for soil nutrients to be replenished.

With **traditional intensive agriculture**, farmers increase their inputs of human and draft animal labor, fertilizer, and water to get a higher yield per area of cultivated land to produce enough food to feed their families and perhaps a surplus for sale. These two forms of traditional agriculture are practiced by about 2.7 billion people—almost half the people on the

Q: How many people have been killed by motor vehicles since the first automobile was built in 1885?

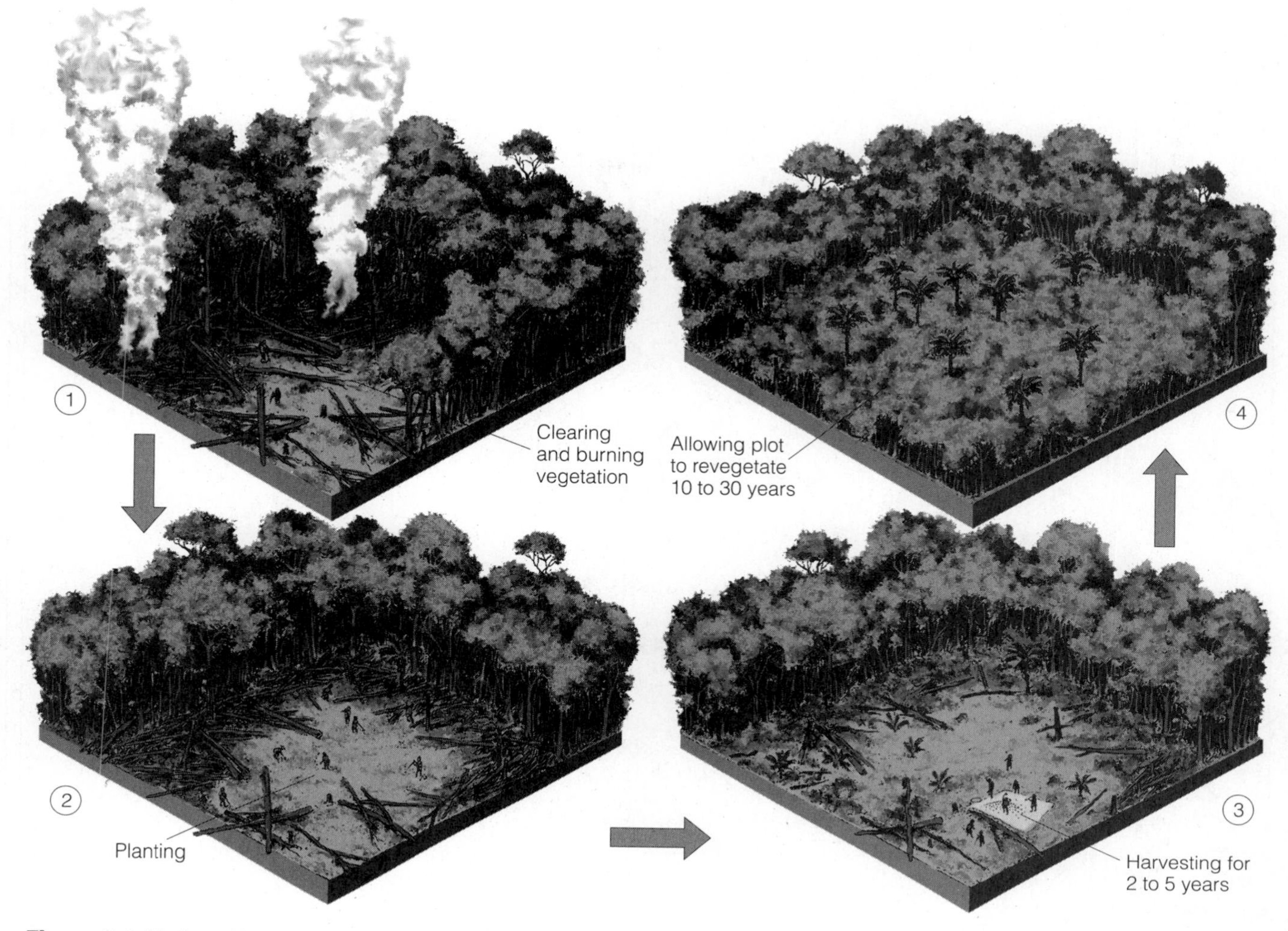

Figure 3-4 Slash-and-burn and shifting cultivation of crops in tropical forests. This method is sustainable if only small plots of the forest are cleared, cultivated for no more than 5 years, and then allowed to lie fallow for 10–30 years to renew soil fertility.

earth—in LDCs. Many traditional farmers simultaneously grow several crops on the same plot, a practice called **interplanting**. Such a use of diversity reduces the chances of losing most or all of their year's food supply to pests or diseases.

INDUSTRIALIZED AGRICULTURE AND GREEN REVOLUTIONS Farmers can produce more food either by farming more land or by getting higher yields per unit of area from existing cropland. Since 1950, most of the increase in global food production has come from raising the yield per unit of land area in a process called a **green revolution**. This process involves planting monocultures of genetically improved plant varieties (often rice or wheat) and lavishing fertilizer, pesticides, and water on them. This approach dramatically increased crop yields in most MDCs between 1950 and 1970 in what is considered the *first green revolution* (Figure 3-5).

Since 1940 U.S. farmers have more than doubled crop production without cultivating more land. They have accomplished this increase in production through industrialized agriculture using green revolution techniques in a favorable climate on some of the world's most fertile and productive soils.

Only 1.8% (4.6 million) of the U.S. population lives on the country's 2.1 million farms, and only about 650,000 Americans work full-time at farming. However, about 23 million people—9% of the population—are involved in the U.S. agricultural system, from growing and processing food to distributing and selling it at the supermarket. In terms of total annual sales, agriculture is the biggest industry in the United States—bigger than the automotive, steel, and housing industries combined. It generates about 18% of the country's gross national product and 19% of all jobs in the private sector, employing more people than any other industry.

A: 18 million (3 million in the United States—twice the number of soldiers killed in all U.S. wars)

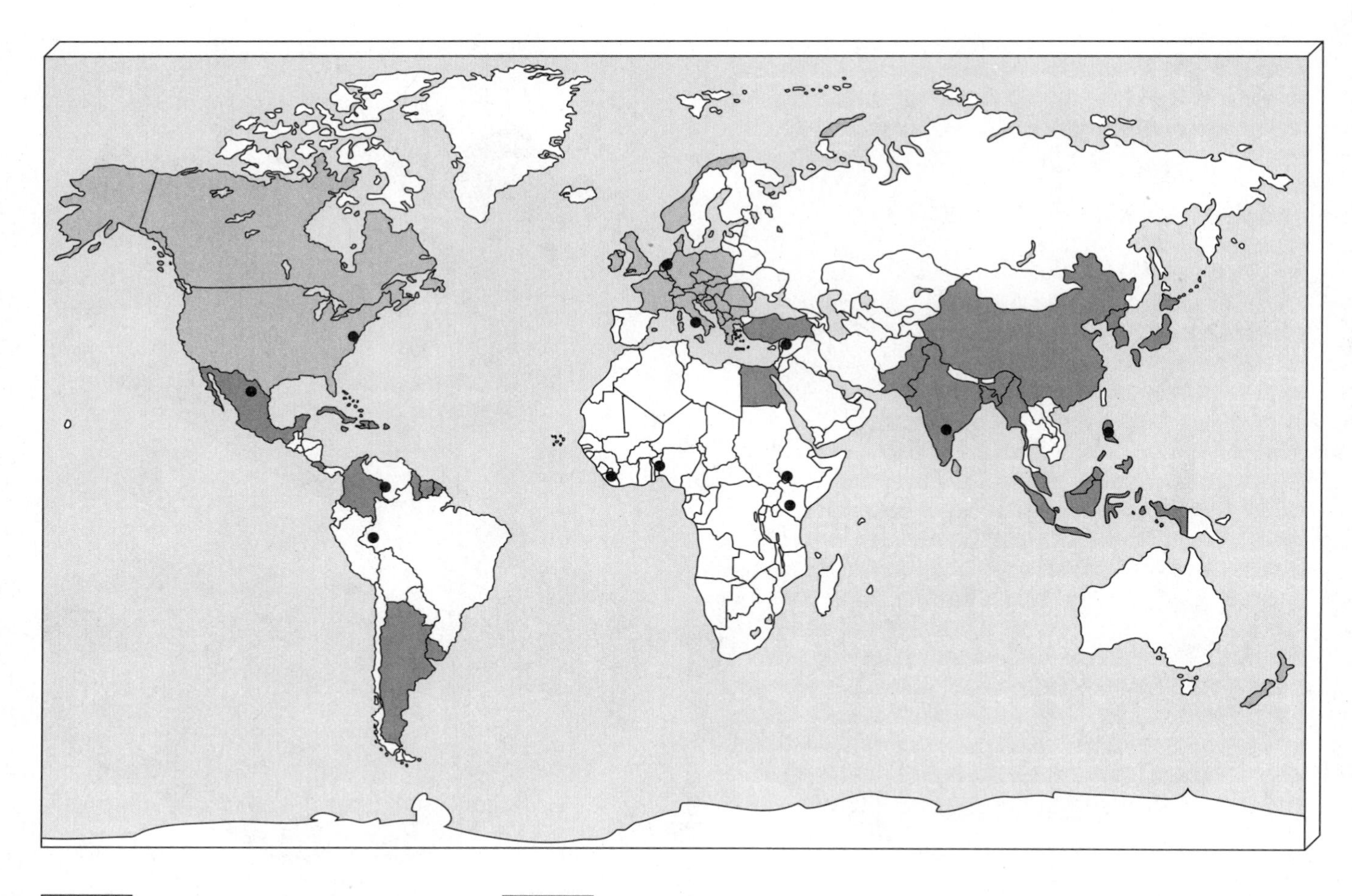

Figure 3-5 Countries whose crop yields per unit of land area increased during the two green revolutions. The first took place in MDCs between 1950 and 1970, and the second occurred since 1967 in LDCs with enough rainfall or irrigation capacity. Thirteen agricultural research centers and genetic storage banks play a key role in developing high-yield crop varieties.

U.S. farmland, called the "breadbasket" of the world, produces half the world's grain exports. In 1991, each U.S. farmer fed and clothed 128 people (94 at home and 34 abroad), up from 58 people in 1976.

Most plant crops in the United States provide more food energy than the energy (mostly from oil and natural gas) used to grow them. However, raising livestock requires much more fossil-fuel energy than the animals provide in food energy.

Energy efficiency is much worse if we look at the whole U.S. food system. Considering the energy used to grow, store, process, package, transport, refrigerate, and cook all plant and animal food, *an average of about 10 units of nonrenewable fossil-fuel energy are needed to put 1 unit of food energy on the table.* By comparison, every unit of energy from the human labor of subsistence farmers provides at least 1 unit of food energy and, with traditional intensive farming, up to 10 units of food energy.

A *second green revolution* has been taking place since 1967, when fast-growing dwarf varieties of wheat and rice, specially bred for tropical and subtropical climates, were introduced into several LDCs (Figure 3-5). With enough fertilizer, water, and pesticides, yields of these new plants can be two to five times those of traditional wheat and rice varieties. And fast growth allows farmers to grow two or even three crops a year (multiple cropping) on the same land parcel. The high inputs of energy, water, fertilizer, and pesticides have yielded dramatic results, but at some point additional inputs become useless because no more output can be squeezed from the land. In fact, yields may even start dropping because the soil erodes, loses fertility, and becomes salty and waterlogged, and because popula-

Q: On what proportion of the world's cropland is soil eroding faster than it forms?

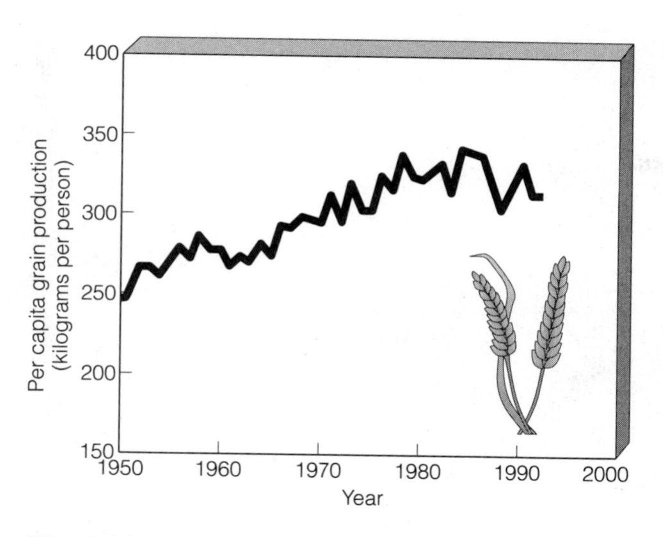

Figure 3-6 World grain production per capita, 1950–1992. (Data from U.S. Department of Agriculture)

tions of rapidly breeding pests develop immunity to widely used pesticides.

3-4 World Food Problems

GOOD NEWS AND BAD NEWS ABOUT FOOD PRODUCTION Between 1950 and 1984, world grain production almost tripled, and per capita production rose by almost 40% (Figure 3-6), reducing hunger and malnutrition around the world. Despite these impressive achievements in total food production, per capita global grain production has declined by roughly 1% per year since 1984 (Figure 3-6). And since 1978, grain production has lagged behind population growth in 69 of the 102 LDCs for which data are available. In 22 African countries, per capita grain production has dropped 28% since 1960 and may drop another 30% during the next 25 years.

Other trends besides population growth reduce per capita availability of food. Since 1978 the irrigated area per person has dropped by 7%; fertilizer use per person has not increased since 1984; and the global fish catch per person has dropped since 1989. Some scientists believe that genetic engineering and other technologies will allow food production to keep up with population growth. Other scientists, however, doubt that this will occur, mostly because of the harmful environmental effects of agriculture. If this is correct, we are faced with two options: reduced food consumption and rising death rates among the population (especially the poor), or an all-out effort to slow population growth.

Saving Children

SOLUTIONS

Officials of the United Nations Children's Fund (UNICEF) estimate that between one-half and two-thirds of childhood deaths from nutrition-related causes could be prevented at an average annual cost of only $5–10 per child—10–19¢ per week. This life-saving program would involve the following simple measures:

- Immunizing children against childhood diseases such as measles
- Encouraging breastfeeding
- Preventing dehydration from diarrhea by giving infants a solution of a fistful of sugar and a pinch of salt in a glass of water
- Preventing blindness by giving people a vitamin A capsule twice a year at a cost of about 75¢ per person
- Providing family-planning services to help mothers space births at least two years apart
- Increasing education for women, with emphasis on nutrition, sterilization of drinking water, and child care

NUTRITIONAL DEFICIENCIES People who cannot grow or buy enough food to meet their basic energy needs suffer from *undernutrition*. To maintain good health and to resist disease, however, people need not only a certain number of calories but also food with the proper amounts of protein, carbohydrates, fats, vitamins, and minerals. People who are forced to live on a low-protein, high-carbohydrate diet consisting only of grains such as wheat, rice, or corn often suffer from **malnutrition**—deficiencies of protein and other key nutrients. Many of the world's desperately poor people (Spotlight, p. 7) suffer from both undernutrition and malnutrition. According to the World Health Organization, about 1.3 billion people—one out of four, and one in three children—are underfed and undernourished (a low estimate is around 0.8 billion). However, saving most children from premature death related to lack of food is possible and not very costly (Solutions, above).

MDCs also have pockets of hunger and malnutrition. For example, a study by Tufts University researchers found that in 1991, roughly 1 out of every 11 Americans and 1 out of 5 American children under age

A: 33%

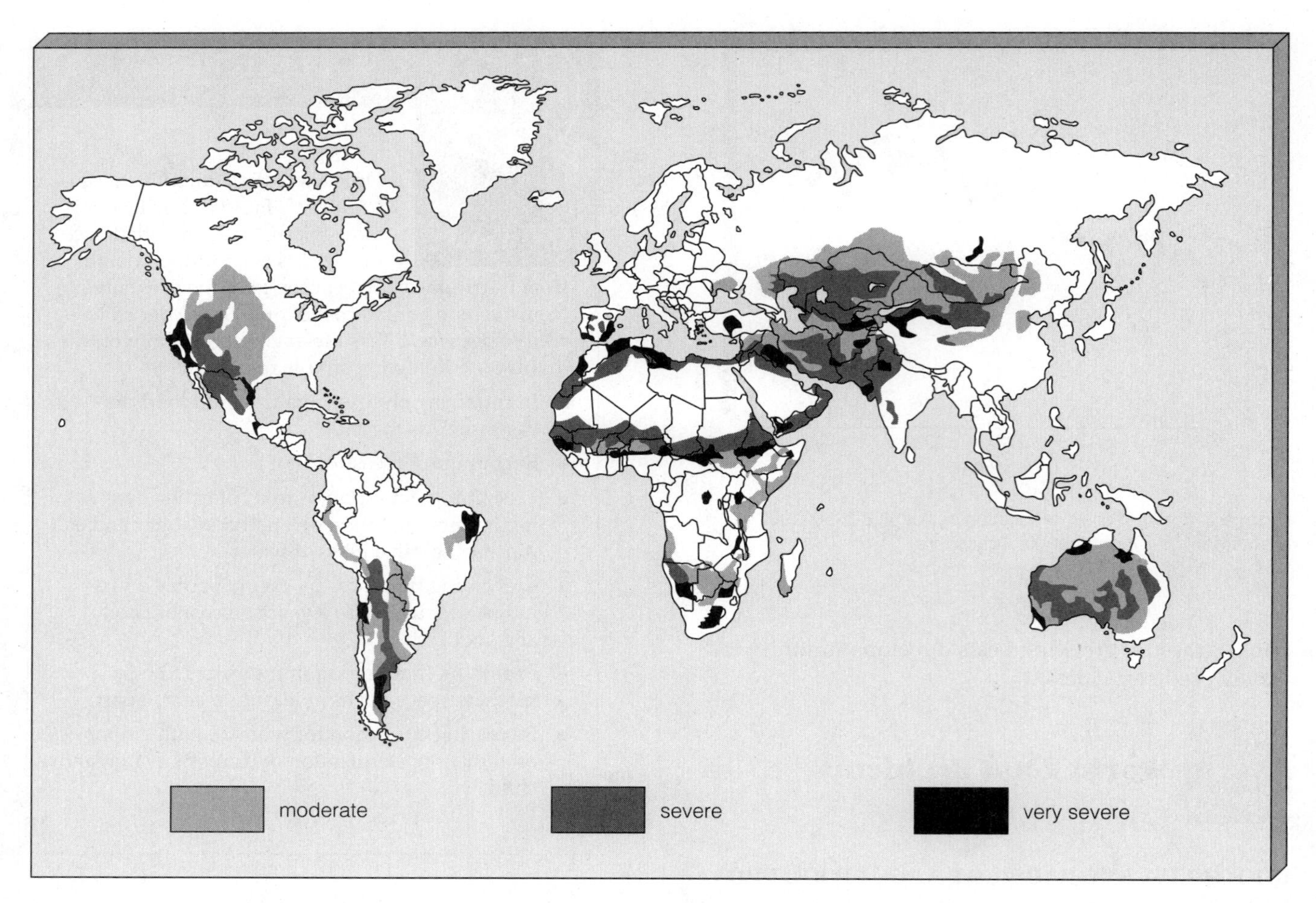

Figure 3-7 Desertification or drops in plant productivity of arid and semiarid lands. The major cause of desertification is overgrazing by livestock. Other causes are prolonged drought, cultivation of land with unsuitable soil, deforestation, and failure to use soil conservation techniques. Worldwide, an area about the size of North and South America combined is threatened by desertification. If current trends continue, desertification could threaten the livelihoods of 1.2 billion people by 2000. (Data from UN Environmental Programme and Harold E. Dregnue)

8 were not getting enough food to prevent undernutrition, malnutrition, or both—mostly because of cuts in food stamps and other forms of government aid for mothers and children between 1980 and 1990.

FOOD SUPPLY AND DISTRIBUTION The good news is that we produce more than enough food to meet the basic needs of every person on Earth. Indeed, if distributed equally, the grain currently produced worldwide would be enough to give 6 billion people—the projected world population in 1998—a meatless subsistence diet. The bad news is that food is not distributed equally among the world's people because of differences in soil, climate, political and economic power, and average per capita income throughout the world.

However, if everyone ate the diet typical of a person in an MDC, with 30–40% of the calories coming from animal products, the current world agricultural system would support only about 2.5 billion people—less than half the present population and only one-fourth of the 10 billion people projected sometime in the twenty-first century (Figure 1-1).

ENVIRONMENTAL EFFECTS OF PRODUCING FOOD Agriculture—both industrialized and traditional—has a greater impact on air, soil, and water resources than any other human activity. These problems include the following:

Soil Degradation

- Erosion.
- Salinization and waterlogging of heavily irrigated soils.

Q: What major food-producing country is doing the most to reduce soil erosion?

- Desertification of arid and semiarid lands, resulting in a decrease in the ability to produce crops or grow natural vegetation (Figure 3-7).

Water Use and Depletion

- Massive use of water to irrigate 18% of the world's cropland.
- Depletion of groundwater by excessive withdrawals for irrigation (Section 7-3).

Pollution

- Air and water pollution from extraction, processing, transportation, and combustion of fossil fuels used in industrialized agriculture (Figure 1-9).
- Air pollution from droplets of pesticides sprayed from planes or ground sprayers.
- Pollution of streams, lakes, and estuaries, and killing of fish and shellfish by pesticide runoff.
- Pollution of groundwater caused by leaching of water-soluble pesticides, nitrates from commercial inorganic fertilizers, and salts from irrigation water.
- Overfertilization of lakes and slow-moving rivers caused by runoff of nitrates and phosphates in commercial inorganic fertilizers, livestock wastes, and food-processing wastes. Livestock in the United States produce 21 times more excrement than is produced by the country's human population. Only about half of this livestock waste is recycled to the soil as organic fertilizer.
- Sediment pollution of surface waters caused by erosion and runoff from farm fields, overgrazed rangeland, deforested land, and animal feedlots.

Loss of Biodiversity

- Loss of genetic diversity in plants caused by clearing biologically diverse grasslands and forests and replacing them with greatly simplified crop fields.
- Endangerment and extinction of wildlife from loss of habitat when grasslands and forests are cleared and wetlands are drained for farming.

Human Health Threats

- Nitrates in drinking water and pesticides in drinking water, food, and the atmosphere.
- Human and animal wastes discharged or washed into irrigation ditches and sources of drinking water.
- Pesticide residues in food.

3-5 Solutions to World Food Problems

INCREASING CROP YIELDS Agricultural experts expect most future increases in food production to come from increased yields per hectare on existing cropland, from improved strains of plants, and from expansion of green-revolution technology to other parts of the world. Scientists are working to create new green revolutions—actually *gene revolutions*—by using genetic engineering and other forms of biotechnology. Over the next 20–40 years they hope to breed high-yield plant strains that are more resistant to insects and disease, thrive on less fertilizer, make their own nitrogen fertilizer (as do legumes), do well in slightly salty soils, can withstand drought, and can use solar energy more efficiently during photosynthesis. Even only occasional innovations could generate enormous increases in global crop production before the middle of the twenty-first century.

However, several factors have limited the success of the green and gene revolutions so far—and may continue to do so:

- Without huge amounts of fertilizer and water, most green-revolution crop varieties produce yields that are no higher (and often are lower) than those from traditional strains; this is the main reason why the second green revolution has not spread to many arid and semiarid areas (Figure 3-5).
- Without ample water, good soil, and favorable weather, new genetically engineered crop strains could fail.
- Continuing to increase inputs of fertilizer, water, and pesticides eventually produces no additional increase in crop yields as the J-shaped curve of crop productivity reaches its limits, levels off, and becomes an S-shaped curve—something that is beginning to happen in some parts of the world.
- Without careful land use and environmental controls, degradation of water and soil can limit the long-term success of green and gene revolutions.
- The cost of genetically engineered crop strains is too high for most of the world's subsistence farmers in LDCs.
- The severe and increasing loss of the earth's biodiversity from deforestation, destruction and degradation of other ecosystems, and replacement of a diverse mixture of natural vegetation with monoculture crops limits the potential of future green and gene revolutions. The genetic

Raising Perennial Crops on the Kansas Prairie

SOLUTIONS

When you think about farms in Kansas, you probably picture endless fields of wheat or corn plowed up and planted each year. By 2040 this picture might change, thanks to pioneering work at the nonprofit Land Institute near Salina, Kansas.

The institute, founded by Wes and Dana Jackson, is experimenting with an ecological approach to agriculture on the midwestern prairie based on planting a mix of *perennial* grasses, legumes, sunflowers, and grain crops in the same field (polyculture); because they are perennial, they don't have to be replanted each year like traditional food crops. The institute's goal is to raise food by mimicking many of the natural processes of the prairie. By eliminating yearly soil preparation and planting, perennial polyculture requires much less labor than growing annual crops.

If the institute and similar groups doing such Earth-sustaining agricultural research succeed, within a few decades, many people may be eating food made from perennials such as *Maximilian sunflower* (which produces seeds with as much protein as soybeans), *eastern gamma grass* (a relative of corn with three times as much protein as corn and twice as much as wheat), *Illinois bundleflower* (a wild nitrogen-producing legume that can enrich the soil and whose seeds may serve as livestock feed), and *giant wild rye* (once eaten by Mongols in Siberia).

uniformity of monoculture increases the vulnerability of food crops to pests and diseases. This, plus widespread species extinction, reduces the genetic raw material needed for future green and gene revolutions. In other words, we are rapidly shrinking the world's genetic "library" just when we need its diversity more than ever.

Wild varieties of the world's most important plants can be collected and stored in gene banks, agricultural research centers (Figure 3-5), and botanical gardens; however, space and money severely limit the number of species that can be preserved there. Moreover, many cannot be stored successfully in gene banks, and power failures, fires, or unintentional disposal of seeds can cause irreversible losses. Further, stored plant species stop evolving and thus are less fit for reintroduction to their native habitats, which may have changed in the meantime.

Because of these limitations, ecologists and plant scientists warn that the only effective way to preserve the genetic diversity of most plant and animal species is to protect representative ecosystems throughout the world from agriculture and other forms of development (Chapter 4).

DEVELOPING NEW FOOD SOURCES Some analysts recommend greatly increased cultivation of various nontraditional plants to supplement or replace such staples as wheat, rice, and corn. One of many possibilities is the winged bean, a protein-rich legume now common only in New Guinea and Southeast Asia. Because of nitrogen-fixing nodules in its roots, this fast-growing plant needs little fertilizer, and it yields so many different edible parts that it has been called a "supermarket on a stalk."

Insects are also important potential sources of protein, vitamins, and minerals. In South Africa, "mopani"—emperor moth larvae—are among several insect food items. Kalahari Desert dwellers eat cockroaches. Lightly toasted butterflies are a favorite food in Bali. French-fried ants are sold on the streets of Bogotá, Colombia, and Malaysians love deep-fried grasshoppers. Most of these insects are 58–78% protein by weight—three to four times as protein-rich as beef, fish, or eggs.

Scientists have identified many plants and insects that could be used as sources of food. The problem is getting farmers to cultivate such crops and insects and convincing consumers to try them.

Some plant scientists believe that we should rely more on mixtures of perennial crops, which are better adapted to regional soil and climate conditions than most annuals, which must be replanted each year (Solutions, above). This strategy would eliminate the need to till soil each year, greatly reducing use of energy; it would also save water and reduce soil erosion and sediment water pollution.

If not overharvested, certain wild animals could be an important source of food. Prolific Amazon river turtles, for example, are used as a source of protein by local people. Another delicacy is the green iguana, "the chicken of the trees." If managed properly, these large, tasty lizards yield up to ten times as much meat as cattle on the same amount of land.

Q: In total sales, what is the biggest industry in the United States?

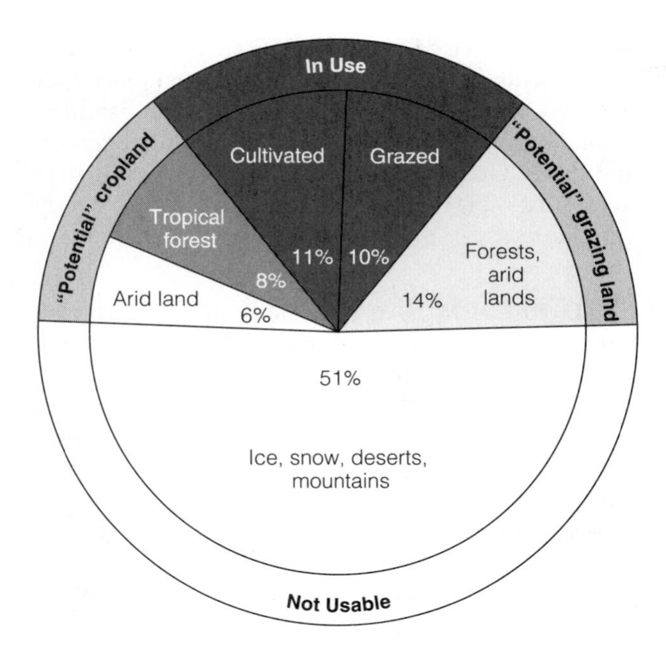

Figure 3-8 Classification of the earth's land. Theoretically, we could double the amount of cropland by clearing tropical forests and irrigating arid lands. However, doing this would destroy valuable forest resources, reduce the earth's biodiversity, and cause other serious environmental problems, usually without being cost-effective.

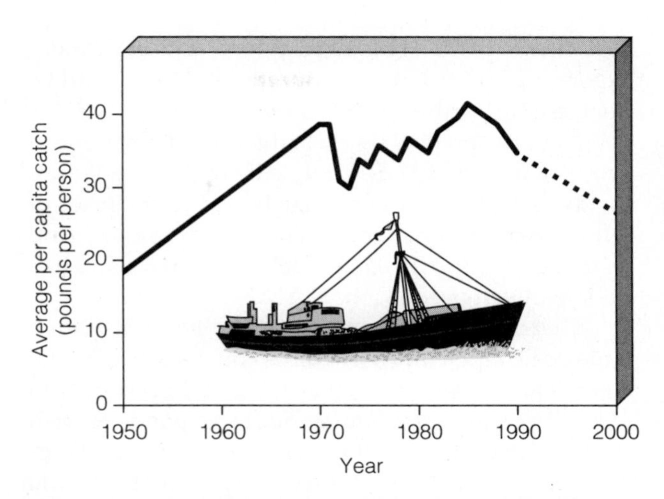

Figure 3-9 Per capita world fish catch has declined since the mid-1980s and is projected to continue dropping through the end of the twentieth century. (Data from United Nations and Worldwatch Institute)

CULTIVATING MORE LAND Theoretically, the world's cropland could be more than doubled by clearing tropical forests and irrigating arid land (Figure 3-8). Clearing tropical forests to grow crops and graze livestock, however, can have disastrous environmental consequences, as is discussed in Section 4-3.

Researchers hope to develop new methods of intensive cultivation in tropical areas. But some scientists argue that it makes more ecological and economic sense to combine the ancient method of shifting cultivation followed by fallow periods that are long enough to restore soil fertility (Figure 3-4) with various forms of interplanting (Figure 3-3d). Scientists also recommend plantation cultivation of rubber trees, oil palms, and banana trees, which are adapted to tropical climates and soils. Such plantations would be established on land that has already been cleared instead of clearing virgin tropical forests.

Much of the world's potentially cultivable land lies in dry areas, especially in Australia and Africa. Irrigating these areas would be very expensive, requiring large inputs of fossil fuel to pump water long distances. In Africa, a large area of potential cropland cannot be used for farming or livestock grazing because it is infested by the tsetse fly. Its bite can infect people with incurable sleeping sickness and can transmit a fatal disease to livestock.

Even if successful, possible increases in cropland in current tropical forests and drylands would not offset the projected loss of almost one-third of today's cultivated cropland from erosion, overgrazing, waterlogging, salinization, mining, and urbanization. And if the global climate changes, as some scientists project, areas where crops can be grown will shift, which will disrupt food production (Section 6-2).

CATCHING MORE FISH Worldwide, people get an average of 20% of their animal protein directly from fish and shellfish, and another 5% indirectly from livestock fed with fish meal. In most Asian coastal and island regions, fish and shellfish supply 30–90% of people's animal protein.

About 86% of the annual commercial catch of fish and shellfish comes from the ocean. Ninety-nine percent of this catch is taken from nutrient-rich and productive waters (mostly estuaries and upwellings) within 370 kilometers (200 nautical miles) of the coast. However, this vital coastal zone is being disrupted and polluted at an alarming rate.

Between 1950 and 1989, the weight of the commercial fish catch increased more than fourfold, but it has declined somewhat in most years since 1989. And since 1970 the per capita catch has declined in most years because the human population has grown at a faster rate than the fish catch (Figure 3-9). Because of overfishing, pollution, and population growth, the per capita world catch is projected to return to the 1960 level by 2000 (Figure 3-9).

RAISING MORE FISH: AQUACULTURE *Aquaculture* can be used to raise fish and shellfish in freshwater ponds or in fenced-in areas or floating cages in coastal lagoons and estuaries. It now supplies about

10% of the world's commercial fish harvest. Fishery experts project that aquaculture production could be doubled during the next 10 years.

Aquaculture has several advantages. First, it can produce high yields. Second, little fuel is needed, so yields and profits are not closely tied to the price of oil, as they are in commercial marine fishing. And third, aquaculture is usually labor-intensive and can provide much-needed jobs in LDCs.

There are problems, however. For one thing, large-scale aquaculture requires considerable capital and scientific knowledge, which are in short supply in LDCs. For another, scooping out huge ponds for fish and shrimp farming in some LDCs has destroyed ecologically important mangrove forests that provide wildlife habitats and help protect shorelines from storm surges and hurricanes. Also, fish in aquaculture ponds can be killed by pesticide runoff from nearby croplands, and bacterial and viral infections of aquatic species can also limit yields. Finally, without adequate pollution control, waste outputs from shrimp farming and other large-scale aquaculture operations can contaminate nearby estuaries, surface water, and groundwater.

INTERNATIONAL FOOD RELIEF: GOOD OR BAD? Most people view food relief as a humanitarian effort to prevent people from dying prematurely. However, some analysts contend that giving food to starving people in countries with high population growth rates does more harm than good in the long run. By not helping people grow their own food and control their population, they argue, food relief can condemn even greater numbers to premature death from starvation and disease in the future.

Biologist Garrett Hardin has suggested that we use the concept of *lifeboat ethics* to decide which countries get food relief. His premise is that there are already too many people in the lifeboat we call the earth. Thus, if food relief is given to countries that are not reducing their populations, this simply adds more people to an already-overcrowded lifeboat. Sooner or later the boat will sink and most of the passengers will drown.

Large amounts of food relief can also depress local food prices, decrease food production, and stimulate mass migration from farms to already-overburdened cities. In addition, food relief discourages local and national governments from investing in rural agricultural development to enable their countries to grow enough food for their populations.

Critics of food relief are not against such aid. Instead, they believe that such aid should be given to help countries control population growth and grow enough food to feed their population by using sustainable agricultural methods. Temporary food relief, they believe, should be given only when there is a complete breakdown of an area's food supply because of natural disaster. Others believe we have a moral obligation to feed the hungry under any circumstances. What do you think?

LAND REFORM An important step in reducing hunger, malnutrition, poverty, and land degradation is land reform. This usually means giving to the landless rural poor in LDCs either ownership or free use of enough previously government-owned land to produce their food and, ideally, enough surplus to provide some income. To date, China and Taiwan have carried out the most extensive land reform programs. The world's most unequal land distribution is in Latin America, where 7% of the population owns 93% of the farmland.

3-6 Protecting Food Resources: Pesticides and Pest Control

PESTICIDES AND THEIR USE A **pest** is any species that competes with us for food, invades lawns, destroys wood in houses, spreads disease, or is simply a nuisance. In diverse ecosystems, populations of species—including the less than 1% we classify as pests—are kept in control by their natural enemies (predators, parasites, and disease organisms)—another crucial type of Earth capital. When we simplify ecosystems, we upset these natural checks and balances that keep any one species from taking over for very long. Then we must devise ways to protect our monoculture crops, tree farms, and lawns from pests that nature once controlled at no charge.

We have done this primarily by developing a variety of **pesticides** (or *biocides*)—chemicals to kill organisms we consider undesirable. Common types of pesticides include *insecticides* (insect-killers), *herbicides* (weed-killers), *fungicides* (fungus-killers), and *rodenticides* (rat- and mouse-killers).

Worldwide, about 0.45 kilograms (1 pound) of pesticides are used yearly for each person on earth. About 75% of these chemicals are used in MDCs, but use in LDCs is soaring. Since 1964, pesticide use in the United States has almost doubled, now amounting to about 4 kilograms (8 pounds) per person per year—about eight times the global average. Most people associate pesticides with crops, but about 40% of U.S.

Q: How many people on average does one U.S. farmer feed?

The Pesticide Treadmill

CONNECTIONS

When genetic resistance to pesticides develops, pesticide sales representatives usually recommend more frequent applications, stronger doses, or a switch to new (usually more expensive) chemicals to keep the resistant species under control. This can put farmers on a **pesticide treadmill**, whereby they pay more and more for a pest control program that does less and less good.

A study by insect ecologist David Pimentel, conducted in 1989 and based on data from more than 300 agricultural scientists and economists, concluded that:

- Although the use of synthetic pesticides has increased 33-fold since 1942, the United States loses more of its crops to pests today (37%) than in the 1940s (31%). Losses attributed to insects almost doubled, from 7% to 13%, despite a 10-fold increase in the use of synthetic insecticides; losses to plant diseases rose from 10% to 12%, and losses to weeds dropped from 14% to 12%.
- The estimated environmental, health, and social costs of pesticide use in the United States range from $4 billion to $10 billion per year.
- Alternative pest control practices could halve the use of chemical pesticides on 40 major crops in the United States without reducing crop yields.
- A 50% cut in pesticide use in the United States would cause retail food prices to rise by only about 0.2% but would raise average income for farmers by about 9%.

lawns are treated with pesticides, typically at levels three to six times higher per acre than are used on farmland.

THE CASE FOR PESTICIDES Proponents of pesticides, who believe that their benefits outweigh their harmful effects, point to the following benefits:

- *Pesticides save human lives.* Since 1945, DDT and an array of other insecticides have probably prevented the premature deaths of at least 7 million people from insect-transmitted diseases such as malaria (carried by the *Anopheles* mosquito), bubonic plague (rat fleas), typhus (body lice and fleas), and sleeping sickness (tsetse fly).
- *Pesticides increase food supplies and lower food costs.* About 55% of the world's potential human food supply is lost to pests before (35%) or after (20%) harvest. In the United States, 37% of the potential food supply is destroyed by pests before and after harvest (13% by insects, 12% by plant pathogens, and 12% by weeds). Without pesticides these losses might be worse, and food prices would rise (by 30–50% in the United States, according to pesticide company officials).
- *Pesticides increase profits for farmers.* Pesticide companies estimate that every $1 spent on pesticides leads to an increase in U.S. crop yields worth approximately $4, but the benefit drops to about $2 if the harmful effects of pesticides are included.
- *Pesticides work faster and better than alternatives.* Pesticides can control most pests quickly and reasonably cheaply, have a long shelf life, are easily shipped and applied, and are safe when handled properly. When genetic resistance occurs, farmers can use stronger doses or switch to other pesticides.
- *The health risks of pesticides are insignificant compared with their health (and other) benefits.* According to Elizabeth Whelan, director of the American Council on Science and Health, which presents the position of the pesticide industry, "the reality is that pesticides, when used in the approved regulatory manner, pose no risk to either farm workers or consumers." Pesticide proponents consider pesticide health-scare news stories to be distorted science and irresponsible reporting, and they point out that about 99.99% of the pesticides we eat are natural chemicals produced by plants to ward off or poison plant-eating organisms.
- *Safer and more effective pesticides are being developed.*

THE CASE AGAINST PESTICIDES Opponents of widespread use of pesticides believe that their harmful effects outweigh their benefits. These harmful effects include the following:

- *Development of genetic resistance to pesticides by pest organisms.* Insects breed rapidly, and within 5–10 years (much sooner in tropical areas) they can develop immunity through natural selection to the chemicals we throw at them (Connections, above). Since 1950 more than 500 major insect pests have developed genetic resistance to one or

A: 128 (94 at home and 34 abroad)

The Day the Roof Fell In

Malaria once infected 9 out of 10 people in North Borneo, now known as Brunei. In 1955 the World Health Organization (WHO) began spraying the island with dieldrin (a DDT relative) to kill malaria-carrying mosquitoes. The program was so successful that the dread disease was virtually eliminated.

Other, unexpected things began to happen, however. The dieldrin killed other insects, including flies and cockroaches living in houses. At first the islanders applauded this turn of events, but then small lizards that also lived in the houses died after gorging themselves on dead insects. Next, cats began dying after feeding on the dead lizards. Without cats, rats flourished and overran the villages, and the people were then threatened by sylvatic plague carried by rat fleas. WHO then parachuted healthy cats onto the island to help control the rats.

Then the villagers' roofs began to fall in. The dieldrin had killed wasps and other insects that fed on a type of caterpillar that either avoided or was not affected by the insecticide. With most of its predators eliminated, the caterpillar population exploded. The larvae munched their way through one of their favorite foods, the leaves used in thatched roofs.

Ultimately, the Borneo episode ended happily: Both malaria and the unexpected effects of the spraying program were brought under control. However, the chain of unforeseen events shows the unpredictability of interfering in an ecosystem.

more insecticides; by 2000 virtually all major insect pest species will probably show some genetic resistance. Because of genetic resistance, most widely used insecticides no longer protect people from insect-transmitted diseases in many parts of the world, leading to even more serious incidences of diseases such as malaria.

- *Killing of natural predators and parasites that keep pest species under control.* With its natural enemies out of the way, a rapidly reproducing insect pest species can make a strong comeback only days or weeks after initially being controlled. Wiping out natural predators can also unleash new pests whose populations the predators had previously held in check and can cause other unexpected effects (Connections, above).
- *Widespread presence in the environment.* No more than 2% (and often less than 0.1%) of the insecticides applied to crops by aerial spraying or ground spraying actually reaches the target pests, and less than 5% of herbicides applied to crops reaches the target weeds. Pesticides that miss their target pests end up in soil, air, surface water, groundwater, bottom sediments in bodies of water, food, and nontarget organisms, including humans and wildlife.
- *Harm to wildlife.* Concentrations of slowly degradable pesticides (such as DDT) can be biologically amplified in food chains and webs (Figure 3-10). The resulting high concentrations of slowly biodegraded, fat-soluble chemicals (such as DDT) can either directly kill the organisms, reduce their ability to reproduce, or make them more vulnerable to diseases, parasites, and predators and thus disrupt food webs. Pesticide runoff from cropland is a leading cause of fish kills worldwide. Use of DDT and related pesticides has been banned in the United States and most MDCs, but these pesticides are still used in some LDCs.
- *Health threats to farm workers.* The World Health Organization estimates that at least 1 million people (including 313,000 farm workers in the United States) are accidentally poisoned by pesticides each year; 4,000–20,000 of them die. At least half of them—and 90% of those killed—are farm workers in LDCs, where educational levels are low, warnings are few, and pesticide regulations are lax or nonexistent. The actual number of pesticide-related illnesses among farm workers in the United States and throughout the world is probably greatly underestimated because of poor records, lack of doctors and disease reporting in rural areas, and faulty diagnoses.
- *Possible health threats to consumers.* According to the U.S. Food and Drug Administration, about 1–3% of the food purchased in the United States has levels of one or more pesticides that are above the legal limit. A 1993 study by the National Academy of Sciences concluded that the legal limits for pesticides in food may need to be reduced by up to 1,000 times to protect children, who are more vulnerable to such chemicals than adults. However, other scientists claim that the

Q: How much of the world's cropland is used to grow livestock feed?

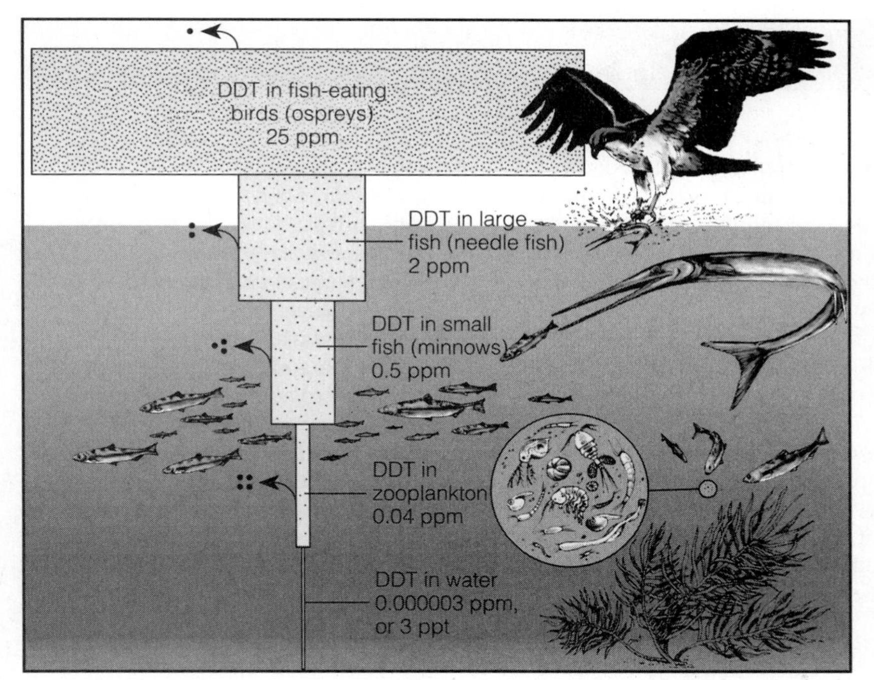

Figure 3-10 DDT concentration in the fatty tissues of organisms was biologically amplified about 10 million times in this food chain in an estuary near Long Island Sound. If each phytoplankton organism in such a food chain retains one unit of DDT from the water, a small fish eating thousands of zooplankton (which feed on the phytoplankton) will store thousands of units of DDT in its fatty tissue. Then a large fish that eats 10 of the smaller fish will receive and store tens of thousands of units, and a bird (or a human) that eats several large fish will ingest hundreds of thousands of units. Dots represent DDT, and arrows show small losses of DDT through respiration and excretion.

risks of harmful effects from pesticide residues on food are overblown. At the same time, other scientists are becoming increasingly concerned about possible genetic mutations, birth defects, nervous system disorders, and effects on the immune and endocrine systems from long-term exposure to low levels of various pesticides. Very little research has been conducted on these effects, and they are not covered adequately in current pesticide, health, and worker safety laws.

- *Inadequate regulation.* According to the National Academy of Sciences, federal laws regulating the use of pesticides in the United States are inadequate and poorly enforced by both the EPA and the Food and Drug Administration. In addition to lobbying for stronger pesticide, health, and worker safety laws, many environmentalists have called for a halt in the export of pesticides banned in the United States to other countries. In what some environmentalists call a *circle of poison*, residues of some of these banned chemicals return to the United States on imported foods (including 25% of the fruits and vegetables consumed each year).

SOLUTIONS: OTHER WAYS TO CONTROL PESTS
Chemicals are not the only answer to pests. Other strategies include the following:

- *Changing the type of crop planted in a field each year* (crop rotation).
- *Planting rows of hedges or trees in and around crop fields.* These plantings hinder insect invasions, provide habitats for their natural enemies, and also reduce erosion of soil by wind (Figure 3-3f).
- *Adjusting planting times so that major insect pests either starve or get eaten by their natural predators.*
- *Growing crops in areas where their major pests do not exist.*
- *Switching from vulnerable monocultures to modernized versions of intercropping that use plant diversity to reduce losses to pests.*
- *Removing diseased or infected plants and stalks and other crop residues that harbor pests.*
- *Using photodegradable plastic to keep weeds from sprouting between crop rows.*
- *Using vacuum machines that gently remove harmful bugs from plants.*
- *Using crossbreeding and genetic engineering to develop food plants and animals that are genetically resistant to certain pest insects, fungi, and diseases.*
- *Using natural enemies* (predators, parasites, and disease-causing bacteria and viruses). More than 300 biological pest control projects worldwide have been successful, especially in China (Solutions, p. 50). In the United States, biological control has saved farmers an average of $25 for every $1 invested. However, biological agents can't always be mass-produced, and farmers find them slower acting and more difficult to apply than

A: More than 50% (66% in the United States)

Along Came a Spider

SOLUTIONS

In 1962, biologist Rachel Carson warned against relying on synthetic chemicals to kill insects and other species we deem pests. Chinese farmers have recently decided that it's time to change strategies. Instead of spraying their rice and cotton fields with poison, they have begun to build little straw huts here and there around the fields in the fall.

If this sounds crazy, it was crazy like a fox. These farmers were giving aid and comfort to insects' worst enemy, one that has hunted them for millions of years: spiders. Protected from the worst of the cold by the huts, far more hibernating spiders would awaken the next spring. Ravenous after their winter fast, they would scuttle off into the fields to stalk their insect prey.

Even without human help, the world's 30,000 known species of spiders kill far more insects every year than insecticides do. The idea of encouraging populations of spiders in fields, forests, and even houses scares most people because spiders have bad reputations. Although a few species of spiders (such as the black widow and brown recluse) are dangerous to people, the vast majority are harmless to humans.

As biologist Thomas Eisner puts it, "Bugs are not going to inherit the earth. They own it now. So we might as well make peace with the landlord." As we seek new ways to coexist with the real rulers of the planet, we would do well to be sure that spiders are in our corner.

pesticides. Also, biological agents must be protected from pesticides sprayed in nearby fields.

- *Using birth control.* Males of some insect pest species (such as the screwworm fly that can infest and kill livestock) can be raised in the laboratory, sterilized by radiation or chemicals, and then released in hordes in an infested area to mate unsuccessfully with fertile wild females. Problems include high costs, difficulties in knowing the mating times and behaviors of each target insect, releasing enough sterile males to do the job, and the relatively few species for which this strategy works.
- *Using insect sex attractants.* In many insect species, a female that is ready to mate releases a minute amount of a chemical sex attractant called a *pheromone*. Pheromones, whether extracted from insects or synthesized in the laboratory, can be used to lure pests into traps or to attract their natural predators into crop fields (usually the more effective approach). These chemicals attract only one species, work in trace amounts, have little chance of causing genetic resistance, and are not harmful to nontarget species. However, it is costly and time-consuming to identify, isolate, and produce the specific sex attractant for each pest or natural predator species.
- *Using insect hormones.* Each step in the insect life cycle is regulated by the timely natural release of certain hormones. These chemicals can either be extracted from insects or synthesized in the laboratory, used to disrupt an insect's normal life cycle, and thus cause the insect to die before it can reach maturity and reproduce. Insect hormones have the same advantages as sex attractants, but they take weeks to kill an insect, are often ineffective with a large infestation of insects, sometimes break down before they can act, and are difficult and costly to produce. They must also be applied at exactly the right time in the target insect's life cycle, and sometimes they can affect the target's predators and other nonpest species.
- *Zapping foods after harvest with gamma radiation.* This strategy can extend the shelf life of some perishable foods and can kill insects, parasitic worms (such as trichinae in pork), and bacteria (such as salmonellae, which infect 51,000 Americans and kill 2,000 each year). According to the FDA and the World Health Organization, over 1,000 studies show that foods exposed to low doses of gamma radiation are safe for human consumption. However, critics argue that it is too soon to see long-term effects (which might not show up for 30–40 years) and that irradiating food destroys some of its vitamins and other nutrients.
- *Using integrated pest management (IPM).* In this approach, each crop and its pests are evaluated as part of an ecological system. A control program is then developed that includes a mix of cultivation and biological and chemical methods used in proper sequence and with the proper timing (Solutions, p. 51).

Integrated Pest Management (IPM)

SOLUTIONS

The overall aim of IPM is not eradication of pest populations, but rather maintenance at just below economically damaging levels. Fields are carefully monitored, and when a damaging level of pests is reached, farmers first use biological and cultivation controls, including vacuuming up harmful bugs. Small amounts of insecticides are applied when absolutely necessary, and the types of chemicals are varied to slow the development of genetic resistance.

In 1986 the Indonesian government banned the use of 57 pesticides on rice and launched a nationwide program to switch to IPM. The results were dramatic: Between 1987 and 1992, pesticide use dropped by 65% and rice production rose by 15%, and the country now saves about $120 million per year on pesticides—enough to cover the cost of its IPM program.

The experiences of countries such as Indonesia, China, Brazil, and the United States have shown that a well-designed IPM program can reduce pesticide use and pest control costs by 50–90%. IPM can also reduce preharvest pest-induced crop losses by 50%, improve crop yields, reduce inputs of fertilizer and irrigation water, and slow the development of genetic resistance because pests are zapped less often and with lower doses of pesticides. Thus IPM is an important form of pollution prevention that reduces risks to wildlife and human health.

However, IPM requires expert knowledge about each pest-crop situation, and it is slower-acting than conventional pesticides. And although long-term costs are typically lower than the costs of using conventional pesticides, initial costs may be higher.

Widespread use of IPM is hindered by government subsidies of conventional chemical pesticides and by opposition from agricultural chemical companies, whose sales would drop sharply. Despite its potential, only about 1% of the U.S. Department of Agriculture's research and education budget is spent on IPM.

Environmentalists urge the USDA to promote integrated pest management by **(1)** adding a 2% sales tax on pesticides to fund IPM research and education; **(2)** setting up a federally supported IPM demonstration project on at least one farm in every county; **(3)** training USDA field personnel and county farm agents in IPM so they can help farmers use this alternative; **(4)** providing federal and state subsidies and perhaps government-backed crop-loss insurance to farmers who use IPM or other approved alternatives to pesticides; and **(5)** gradually phasing out subsidies to farmers who depend almost entirely on pesticides, once effective IPM methods have been developed for major pest species.

3-7 Solutions: Sustainable Food Production

GUIDELINES FOR SUSTAINABLE AGRICULTURAL SYSTEMS To many environmentalists, the key to reducing world hunger, poverty, and the harmful environmental impacts of both industrialized and traditional agriculture is to develop a variety of **sustainable agricultural systems**. This approach involves combining the wisdom of traditional agricultural systems with new techniques that take advantage of local climates, soils, resources, and cultural systems. Here are some general guidelines:

- *Neither rob the soil of nutrients nor waste water, and return whatever is taken from the earth.* This means minimizing soil erosion by a variety of methods (Figure 3-3), not cultivating easily erodible land, preventing overgrazing by livestock, and using organic fertilizers, crop rotation, and intercropping to increase the organic content of soils. It also means ceasing attempts to grow water-thirsty crops in arid and semiarid areas, raising water prices to encourage water conservation, and using irrigation systems that minimize water waste, salinization, and waterlogging.
- *Encourage systems featuring a diverse mix of crops and livestock, instead of depending too much on production of a single crop or livestock type.*
- *Rely as much as possible on locally available, renewable biological resources and use them in ways that preserve their renewability.* Examples include using organic fertilizers from animal and crop wastes (green manure and compost); planting fast-growing trees to supply fuelwood and add nitrogen to

A: 1.3 billion—one person of every four on Earth

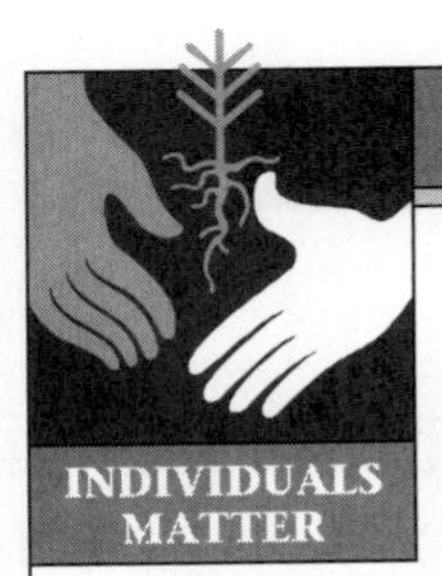

What *You* Can Do to Promote Sustainable Agriculture

- *Waste less food.* An estimated 25% of all food produced in the United States is wasted; it rots in the supermarket or refrigerator, or it is scraped off the plate and into the garbage in households and restaurants.
- *Eat lower on the food chain.* This can be done by reducing or eliminating meat consumption to reduce environmental impact and improve human health.
- *Don't feed your dog or cat canned meat products.* Balanced-grain pet foods are available and are better for your pet.
- *Help reduce the use of pesticides on agricultural products by asking grocery stores to stock (and by buying) fresh produce and meat produced by organic methods.* About 0.5% of U.S. farmers grow about 3% of the country's crops using organic methods.
- *Grow some of your own food using organic farming techniques and drip irrigation to water your crops.*
- *Give up the idea that "the only good bug is a dead bug" ("bugitis").* Recognize that full-scale chemical warfare on insect pests wipes out many of our insect allies.
- *Don't insist on perfect-looking fruits and vegetables.* Such produce is more likely to contain high levels of pesticide residues.
- *Use pesticides around your home only when absolutely necessary, and use them in the smallest amount possible.*
- *Think globally, eat locally.* Whenever possible, eat food that is locally grown and in season. This supports your local economy, gives you more influence over how the food is grown (organic vs. conventional methods), saves energy used to transport food over long distances, and reduces the use of fossil fuels and pollution. If you deal directly with local farmers, you can also save money.
- *Pressure elected officials to develop and encourage sustainable agriculture in the United States and throughout the world.*

the soil; building simple devices for capturing and storing rainwater for irrigating crops; and cultivating crops adapted to local growing conditions.

- *Greatly reduce the use of fossil fuels in agriculture by increasing use of locally available renewable energy resources such as sun, wind, and flowing water, and by using more organic fertilizer instead of commercial inorganic fertilizer.*
- *Emphasize biological pest control and integrated pest management (Solutions, p. 51) instead of overuse of chemical pesticides.*
- *Provide economic incentives for farmers using sustainable agriculture.*

MAKING THE TRANSITION TO SUSTAINABLE AGRICULTURE IN MDCS In MDCs such as the United States, a shift to sustainable agriculture will not be easy, for it will be opposed by agribusiness, by successful farmers with large investments in unsustainable forms of industrialized agriculture, and by specialized farmers unwilling to learn the demanding art of farming sustainably. However, environmentalists believe that this shift could be brought about over a 10- to 20-year period by the following means:

- *Greatly increase government support of research and development of sustainable agricultural methods and equipment.* At present, only about 1% of the Department of Agriculture's annual research budget is used for this purpose.
- *Set up demonstration projects in each county so that farmers can see how sustainable systems work.*
- *Establish training programs in sustainable agriculture for farmers, county farm agents, and Department of Agriculture personnel.*
- *Establish college curricula for sustainable agriculture.*
- *Give subsidies and tax breaks to farmers using sustainable agriculture and to agribusiness companies that develop products for this type of farming.* For example, Iowa taxes fertilizers and pesticides; it then uses the revenues to support research into and development of sustainable agriculture. Minnesota provides low-interest loans for farmers engaged in sustainable agriculture. Austria, Denmark, Finland, Germany, Norway, and Sweden offer 3–5-year subsidies for farmers converting to sustainable agriculture.

Each of us can play a part in greatly increasing the use of sustainable agriculture at the local, national, and global levels (Individuals Matter, above).

Q: How many people die each year from hunger-related causes?

At some point, either the loss of topsoil from the world's croplands will have to be checked by effective soil conservation practices, or the growth in the world's population will be checked by hunger and malnutrition.

LESTER R. BROWN

CRITICAL THINKING

1. Why should everyone, not just farmers, be concerned with soil conservation?
2. What are the main advantages and disadvantages of commercial inorganic fertilizers? Why should both inorganic and organic fertilizers be used?
3. Summarize the advantages and limitations of each of the following proposals for increasing world food supplies and reducing hunger over the next 30 years: **(a)** cultivating more land by clearing tropical forests and irrigating arid lands, **(b)** catching more fish in the open sea, **(c)** producing more fish and shellfish through aquaculture, and **(d)** increasing the yield per unit area of cropland.
4. Is sending food to famine victims helpful or harmful? Explain. Are there any conditions you would attach to sending such relief? Explain.
5. Should tax breaks and subsidies be used to encourage more U.S. farmers to switch to sustainable agriculture and integrated pest management? Explain.
6. Should U.S. companies be allowed to export to other countries pesticides that have been banned or severely restricted in the United States? Explain.

A: 40 million (some say 20 million and others say 60 million)

4 Biodiversity: Sustaining Ecosystems and Species

Forests precede civilizations, deserts follow them.

FRANCOIS-AUGUSTE-RENÉ DE CHATEAUBRIAND

4-1 The Importance of Ecological Diversity

SUSTAINING ECOSYSTEMS Forests, grasslands, deserts, wetlands, coral reefs, and other ecosystems throughout the world are coming under increasing stress from population growth and economic development. In Chapter 1 you learned that there are three components of the planet's biodiversity: **(1)** *genetic diversity*—variability in the genetic makeup among individuals within a single species, **(2)** *species diversity*—the variety of species on Earth and in different habitats of the planet, and **(3)** *ecological diversity*—the variety of forests, deserts, grasslands, streams, lakes, oceans, and other biological communities that interact with one another and with their nonliving environments.

Because biodiversity is a vital part of the Earth capital that sustains all life, preserving the planet's genes, species, and ecosystems should be among our most important priorities. One way to do this is to protect species from sharp population declines and premature extinctions that result from human activities. However, most wildlife biologists believe that the best way to protect species diversity is to sustain and protect the earth's ecosystems that serve as habitats. This means establishing a worldwide network of reserves, parks, wildlife sanctuaries, and other protected areas. Protecting these vital oases of biodiversity from damage, using them sustainably by learning how nature does this, and helping heal those we have damaged are important challenges.

This scientific approach recognizes that saving wildlife species means saving the places where they live. It is also based on Aldo Leopold's ethical principle that something is right when it tends to maintain Earth's life-support systems for us and other species, and wrong when it doesn't.

As of 1993 there were about 7,000 protected areas throughout the world, occupying 4.9% of the earth's land surface. That is an important beginning, but environmentalists say that a minimum of 10% of the globe's land area must be protected. Moreover, many existing reserves are too small to provide any real protection for the populations of wild species that live on them. And many types of ecosystems are not included.

An international fund to help LDCs protect and manage 10% of the planet's land area would cost $100 million per year—about what the world's nations spend on arms every 90 minutes. Because there won't be enough money to protect most of the world's biodiversity, conservation biologists believe that efforts should be focused on the so-called megadiversity countries (Figure 4-1).

PUBLIC LANDS IN THE UNITED STATES No nation has set aside so much of its land—about 42%—for public use, resource extraction, enjoyment, and wildlife as the United States. Almost one-third of the country's land is managed by the federal government; 73% of this public land is in Alaska, and another 22% is in the western states. These public lands—owned jointly by all U.S. citizens—include: **(1)** 156 *national forests* and 19 *national grasslands*, managed by the Forest Service; **(2)** *national resource lands* (mostly grasslands and deserts), managed by the Bureau of Land Management; **(3)** 503 *national wildlife refuges*, managed by the Fish and Wildlife Service; **(4)** 367 units in the *national park system* (including 50 major parks), overseen by the National Park Service; and **(5)** the *National Wilderness Preservation System*, consisting of 474 roadless areas within the national parks, national wildlife refuges, and national forests that are protected from development.

HOW SHOULD PUBLIC LANDS BE MANAGED? Since 1901 conservationists have been split into two major schools of thought on how U.S. public lands should be used and managed. *Preservationists* have sought to protect large areas from mining, logging, and other forms of resource extraction so that these lands can be enjoyed today and passed on unspoiled to future generations; mostly they have fought a losing battle. Members of the *wise-use* school see public lands as resources to be used wisely to enhance eco-

Q: What human activity has the most harmful overall environmental impact?

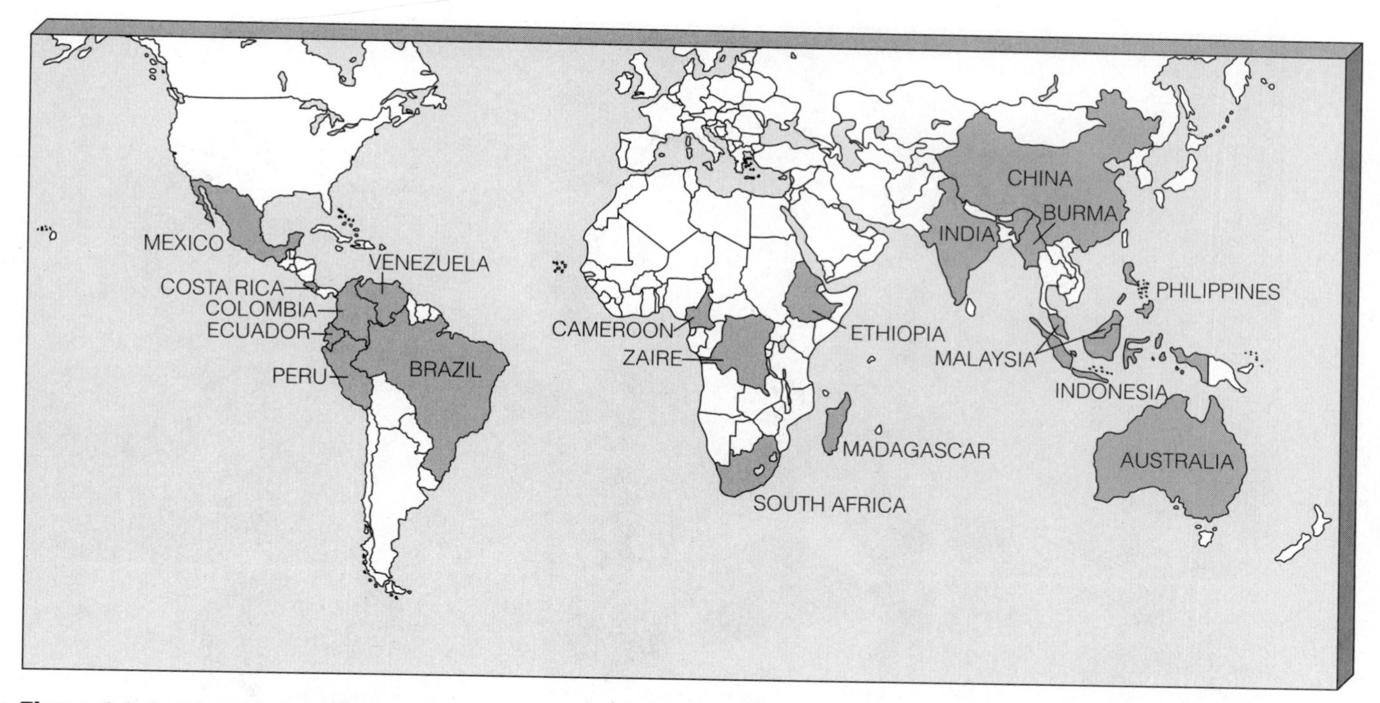

Figure 4-1 Earth's megadiversity countries. Conservation biologists believe that efforts to preserve repositories of biodiversity should be concentrated in these species-rich countries. (Data from Conservation International and the World Wildlife Fund)

nomic growth and national strength; they advocate efficient and scientific management of these lands to produce sustainable yields of potentially renewable resources (such as trees, and grasses for livestock grazing) and to enable extraction of mineral and energy resources.

Despite their basic differences, both schools of early conservationists opposed delivering these public resources into the hands of the few for profit. Both groups have been disappointed. Since 1910 development rights to public lands have routinely been sold at below market prices to large corporate farms, ranches, mining companies, and timber companies. Taxpayers have subsidized this use of resources by absorbing the loss of potential revenue and paying for most of the resulting damage.

Since the early 1900s a third group of people has attempted to convince the federal government to give or sell most of the nation's public lands to states or to private interests so they can be more fully exploited for economic growth. In the early 1930s, President Herbert Hoover supported this policy, but the Great Depression of the 1930s made the cost of owning such lands unattractive to state governments and private investors.

In the late 1970s a coalition of ranchers, miners, loggers, developers, farmers, politicians, and others launched a political campaign known as the *sagebrush rebellion* with the goal of turning over most western public lands to the states or to private interests. This idea was strongly supported by President Ronald Reagan (a declared "sagebrush rebel") in the 1980s, but was thwarted by strong opposition in Congress, public outrage, and legal challenges by environmental and conservation organizations, whose memberships soared in this period.

In 1988, several hundred local and regional grassroots groups (many financed mostly by developers, timber, mining, oil, coal, and ranching interests) formed a national coalition with the goals of destroying the environmental movement in the United States, increasing resource exploitation of public lands at low prices, and turning most of these lands over to private enterprise. They try to confuse the public by calling themselves the "wise-use movement." Some conservationists (including members of the preservationist and original wise-use schools of thought), however, call them the "Earth and people abuse movement."

4-2 Sustaining and Managing Forests

TYPES OF FORESTS **Old-growth forests** are virgin (uncut) forests and old second-growth forests that have not been seriously disturbed for several hundred

Figure 4-2 Stratification of specialized plant and animal niches in various layers of an old-growth tropical rain forest. The presence of specialized niches allows species to avoid or minimize competition for resources and results in the coexistence of a great variety of species (biodiversity). Niche specialization is promoted by the adaptation of plants to the different levels of light available in the forest's layers and by hundreds of thousands of years of evolution in a fairly constant climate.

years. They contain massive trees that are hundreds or even thousands of years old. The understory and other vegetation zones in old-growth forests provide ecological niches for a variety of wildlife species (Figure 4-2). These forests also have large numbers of standing dead trees (snags) and fallen logs (boles), which are habitats for a variety of species. Decay of this dead vegetation returns nutrients to the soil.

Second-growth forests are stands of trees resulting from secondary ecological succession after cutting (Figure 4-3). Most forests in the United States and other temperate regions are second-growth forests that grew back after virgin forests were logged or farms were abandoned. About 40% of tropical forests are second-growth forests. Some old second-growth stands have remained undisturbed long enough to be classified as old-growth forests, but many are *tree farms*—managed tracts of uniformly aged trees of one species that are harvested as soon as they become commercially valuable.

COMMERCIAL AND ECOLOGICAL IMPORTANCE OF FORESTS Forests give us lumber for housing, biomass for fuelwood, pulp for paper, medicines, and many other valuable products such as turpentine, nuts and fruit, and charcoal. Many forestlands are also used for mining, grazing livestock, and recreation.

Forested watersheds act as giant sponges, slowing down runoff and absorbing and holding water that recharges springs, streams, and groundwater.

Q: What percentage of the world's potential food supply is lost to pests?

Figure 4-3 Secondary ecological succession of plant communities on a farm field in North Carolina that was cleared of an old-growth forest, planted in crops for a number of years, and then abandoned because of soil depletion. After the farmland was abandoned it took about 150 years for the area to be covered again with a second-growth oak and hickory forest.

Thus they regulate the flow of water from mountain highlands to croplands and urban areas, and they help control soil erosion, reduce flooding, and reduce the amount of sediment washing into streams, lakes, and reservoirs.

Forests also influence local, regional, and global climate. For example, 50–80% of the moisture in the air above tropical forests (Figure 4-2) comes from trees via transpiration and evaporation. If large areas of these lush forests are cleared, average annual precipitation drops, the region's climate gets hotter and drier, and soils become depleted of already-scarce nutrients, baked, and washed away. Eventually this process can convert a diverse tropical forest into a sparse grassland or even a desert.

Forests also provide habitats for more wildlife species than any other biome, making them the planet's major reservoir of biodiversity. They also buffer us against noise, absorb air pollutants, and nourish the human spirit. Since agriculture began about 10,000 years ago, human activities have reduced the earth's forest cover by at least one-third, to about 34% of the world's land area.

FOREST MANAGEMENT IN THE UNITED STATES

Though forests cover about one-third of the lower 48 states, most of the old-growth forests there have been cut (Figure 4-4), and most of what remains is threatened. About 22% of the commercial forest acreage in the United States is located within the 156 national forests managed by the U.S. Forest Service. These forestlands serve as grazing lands for more than 3 million cattle and sheep each year, support multimillion-dollar mining operations, contain a network of roads eight times longer than the entire U.S. interstate highway system, and receive more recreational visits than any other federal public lands.

A: 55% (35% before harvest and 20% after harvest)

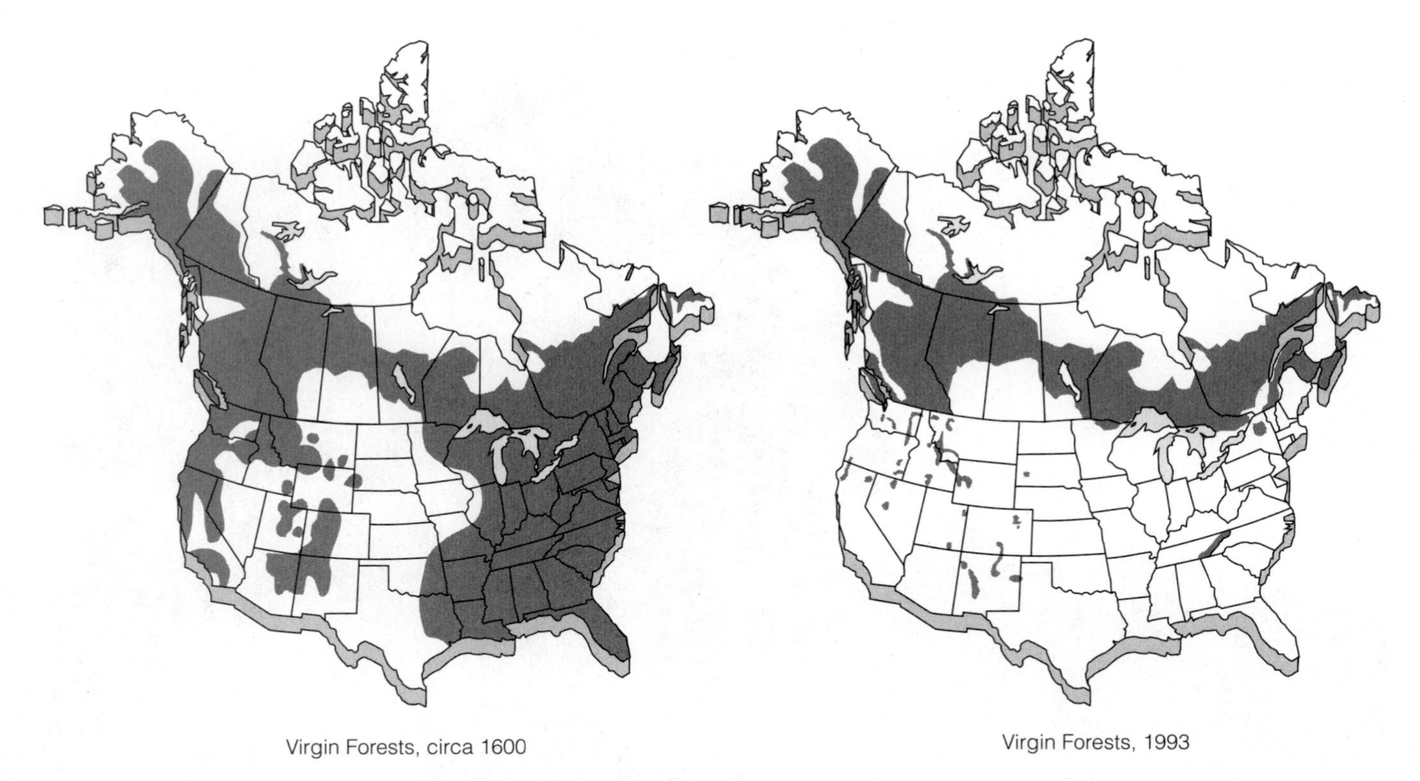

Figure 4-4 Vanishing old-growth forests in the United States and Canada. Since about 1600, 90–95% of the virgin forests that once covered much of the lower 48 states have been cleared away; most of the remaining old-growth forests in the lower 48 states and Alaska are on public lands. About 60% of old-growth forests in western Canada have been cleared, and much of what remains is slated for cutting. (Data from the Wilderness Society and the U.S. Forest Service, and *Atlas Historique du Canada*, Vol. 1)

There is much controversy over how national forests should be used. Timber company officials complain that they aren't allowed to buy and cut enough timber on public lands, especially in remaining old-growth forests in California and the Pacific Northwest (Figure 4-4).

By law, the Forest Service is supposed to manage national forests according to a *multiple use* philosophy; that is, national forests are to be used for a variety of purposes, including wildlife conservation, recreation, grazing, and timber harvesting. However, environmentalists charge that since 1960 the Forest Service has increasingly allowed timber harvesting to become the dominant use in most national forests. They point out that almost three-fourths of the Forest Service budget is devoted directly or indirectly to the sale of timber, and that at current rates of timber removal—the equivalent of about 129 football fields a day—all unprotected old-growth forests on public lands in western Washington and Oregon will be gone by the year 2023.

Because the Forest Service keeps most of the money it makes on timber sales, it has a powerful built-in incentive to encourage timber sales. Local county commissioners also exert tremendous pressure on both members of Congress and Forest Service officials to keep timber harvests high because counties get 25% of the gross receipts from national forests within their boundaries.

Environmentalists and the U.S. General Accounting Office have accused the Forest Service of poor fiscal management of public forests. By law, the Forest Service must sell timber for no less than the cost of reforesting the land from which it was harvested. However, the cost of access roads is not included in this price and is provided as a subsidy to logging companies. Logging companies also get the timber itself for less than they would normally pay a private landowner.

Studies have shown that between 1978 and 1992, national forests lost at least $4.2 billion (some sources say $7 billion) from timber sales. With interest, this loss added at least $5.9 billion to the national debt. Timber company officials, however, argue that being able to get timber from federal lands fairly cheaply benefits taxpayers by keeping lumber prices down.

Q: What percentage of U.S. crops are lost to pests?

THE CONTROVERSY OVER OLD-GROWTH DEFORESTATION IN THE PACIFIC NORTHWEST To officials of timber companies, the giant living trees and rotting dead trees in old-growth forests are valuable resources that should be harvested for profit and to provide jobs, not locked up to please environmentalists. They point out that the timber industry annually pumps millions of dollars into the Pacific Northwest's economy and provides jobs for about 100,000 loggers and millworkers.

To environmentalists, the remaining ancient forests on the nation's public lands are a treasure whose ecological, scientific, aesthetic, and recreational values far exceed the economic value of cutting them down for short-term economic gain. Rotting trees, for example, are a major way nutrients are returned to the soil to support existing and new plant growth. The fate of these forests is a national issue because these forests are owned by all U.S. citizens, not just the timber industry or the residents of a region. It is also a global issue because these forests are important reservoirs of irreplaceable biodiversity, and because the way the United States treats its few remaining old-growth forests sets a precedent for other nations' (especially LDCs') treatment of their old-growth forests, wetlands, coral reefs, and other ecosystems.

The threatened northern spotted owl, which lives in many of these forests, has become a symbol in the struggle between environmentalists and timber company officials over the fate of unprotected old-growth forests on public lands in the Pacific Northwest. Despite simplistic media coverage and speeches by politicians, the controversy over cutting of ancient forests in the Pacific Northwest isn't an owl-versus-jobs issue. Timber jobs are disappearing in the Northwest for the same reasons the owls are: The ancient forests they depend on are almost gone.

Loggers, millworkers, and store owners who live in these communities are caught in the middle, pawns in a high-stakes game of corporate profit. They correctly fear for their jobs, but automation, export of raw logs (mostly to Japan), and cutting of most remaining old-growth trees will also do away with their jobs.

Most environmentalists believe that supporting sustainable use of public forests—based on allowing limited selective cutting, replanting and restoring cleared areas, diversifying the economy, and encouraging tourism—is the best way logging-based communities can remain economically and ecologically healthy.

Another part of the solution to this dilemma is to recognize that we are all part of the problem. We buy wood that is harvested from old-growth forests at such a low price that sustainable logging is not economically feasible. The marketplace is not indicating the real costs of destroying and degrading our forests because we don't insist that the prices of wood and wood products include their full short- and long-term environmental and social costs. Until we change the market system to include these real costs (Solutions, p. 5), we have a powerful built-in economic incentive to deplete potentially sustainable Earth capital and eliminate potentially sustainable jobs.

SOLUTIONS: REFORMING FEDERAL FOREST MANAGEMENT Forestry experts and environmentalists have suggested several ways to reduce overexploitation of publicly owned timber resources, including the following:

- *Urge elected representatives to ban all timber cutting in national forests and to fund the Forest Service completely from recreational user fees.* The Forest Service estimates that recreational user fees would generate three times what it earns from timber sales.
- *Until a total ban is enacted, reduce the current annual harvest of timber from national forests by one-half,* instead of doubling it as proposed by the timber industry.
- *Preserve at least 50% of remaining old-growth timber in any national forest.*
- *Allow individuals or groups to buy conservation easements that prevent timber harvesting on designated areas of public old-growth forests.*
- *Build no more roads in national forests.*
- *Require that timber from national forests be sold at a price that includes the costs of road building, site preparation, and site regeneration, and strictly enforce the existing law requiring that all timber sales in national forests yield a profit for taxpayers.*
- *Don't use money from timber sales in national forests to supplement the Forest Service budget,* which encourages overexploitation of timber resources.
- *Eliminate the provision that returns 25% of gross receipts from national forests to counties containing the forests, or base such returns on recreational user-fee receipts only.*
- *Provide federal funds for restoration and reforestation of cut and degraded areas of national forests* to help renew these ecosystems and furnish alternative jobs for unemployed loggers and millworkers.

Timber company officials vigorously oppose most of these proposals, claiming they would cause economic disruption in their industry and in logging communities and raise the price of timber for consumers. Environmentalists argue that taxpayers are paying higher prices than they think for timber when their tax dollars are used to subsidize logging in national

A: 37% (7% higher than in the 1940s despite a 33-fold increase in pesticide use since then)

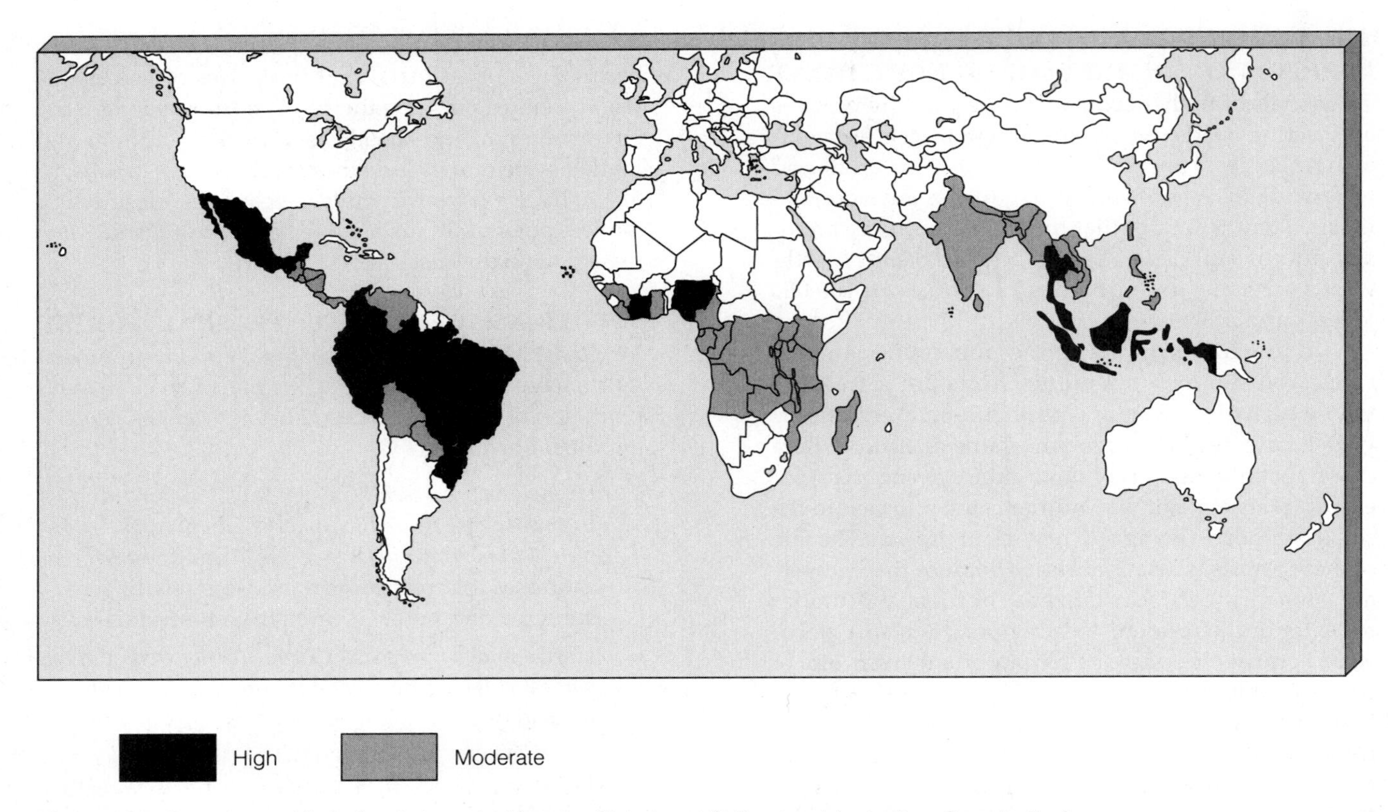

Figure 4-5 Countries rapidly losing their tropical forests. (Data from UN Food and Agriculture Organization)

forests. They contend that including these and the harmful environmental costs of unsustainable timber cutting would promote more sustainable use of these resources, help sustain logging communities, and protect biodiversity.

4-3 Combating Tropical Deforestation

THE LOSS OF TROPICAL FORESTS Tropical forests, which cover about 6% of the earth's land area, grow near the equator in Latin America, Africa, and Asia. About 56% of the world's tropical forests have already been cleared or damaged. Satellite scans and ground-level surveys indicate that the remaining forests are vanishing rapidly (Figure 4-5), at a rate of at least 154,000 square kilometers (59,000 square miles) per year—equivalent to about 34 city blocks per minute, or almost two football fields per second. It's estimated that an equivalent area of these forests is damaged every year.

Reforestation in the tropics scarcely deserves the name, with only one tree planted for every 10 trees cut; in Africa the rate is 1 to 29. If the current rate of loss continues, all remaining tropical forests (except for a few preserved but still vulnerable patches) will be gone within 30 to 50 years, and much sooner in some areas.

WHY SHOULD WE CARE ABOUT TROPICAL FORESTS? Environmentalists consider the plight of tropical forests to be one of the world's most serious environmental problems. These forests are home to at least 50% (some estimate 90%) of the earth's total stock of species—most of them still unknown and unnamed.

Tropical forests touch the daily lives of everyone on Earth through the products and ecological services they provide. These forests supply half of the world's annual harvest of hardwood, hundreds of food products (including coffee, tea, cocoa, spices, nuts, chocolate, and tropical fruits), and many materials (including natural latex rubber, resins, dyes, and essential oils) that can be harvested sustainably and generate twice as much revenue per hectare as timber production and three times as much as cattle ranching.

The active ingredients for 25% of the world's prescription drugs are substances derived from plants, most of which grow in tropical rain forests. Such drugs are used in birth control pills, tranquilizers, muscle relaxers, and life-saving drugs for treating malaria,

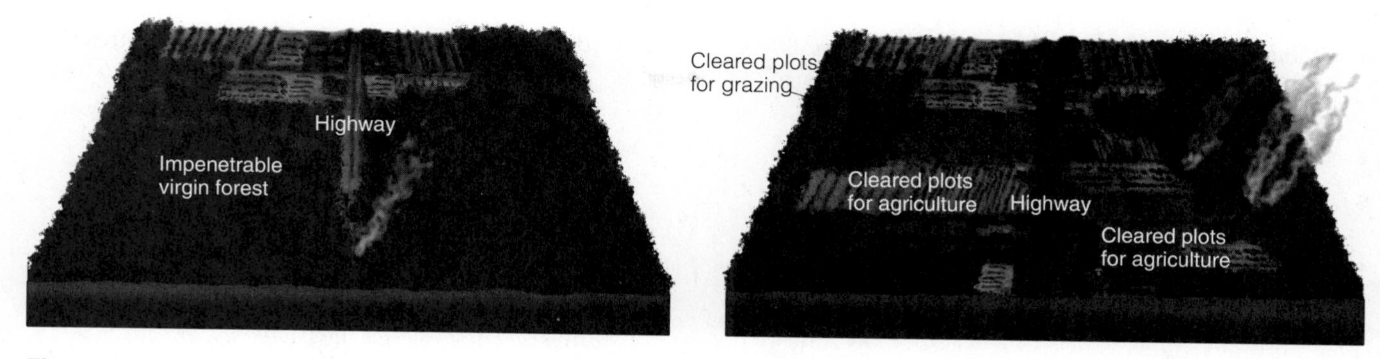

Figure 4-6 Building roads into previously inaccessible forests paves the way for timber harvesting, but also to degradation and destruction. Roads open up once-impenetrable forests to farmers, miners, ranchers, and dam builders who can damage, harvest, or flood large areas of trees, and to hunters who can deplete some wild animal species.

leukemia and Hodgkin's disease, testicular and lung cancer, heart disease, high blood pressure, multiple sclerosis, venereal warts, and many other diseases. Seventy percent of the 3,000 plants identified by the National Cancer Institute as sources of cancer-fighting chemicals come from tropical forests. While you are reading this page, a plant species that could cure a type of cancer, AIDS, or some other deadly disease might be wiped out forever.

Botanists also believe that tens of thousands of strains of plants with potential food value await discovery in tropical forests. Despite their immense potential, less than 1% of the estimated flowering plant species in the world's tropical forests have been examined closely for their possible use as human resources.

Biologist E. O. Wilson warns that destroying these forests and the species they support for short-term economic gain is like throwing away a wrapped present or burning down an ancient library before you read the books. In addition, the Environmental Policy Institute estimates that unless destruction of tropical forests stops, the resulting flooding and loss of topsoil could cause as many as a billion people to starve during the next 30 years.

CAUSES OF TROPICAL DEFORESTATION The two major underlying causes of the current massive destruction and degradation of tropical forests are:

- *Population growth and poverty*, which combine to drive subsistence farmers and the landless poor to tropical forests to try to grow enough food to survive, often using unsustainable methods.
- *Massive foreign debt and policies of governments and international development and lending agencies that encourage rapid depletion of resources to stimulate short-term economic growth.* LDCs are encouraged to borrow huge sums of money from MDCs to finance economic growth. To pay the interest on their debts, these countries often sell off their forests, minerals, oil, and other resources—mostly to MDCs—at low prices dictated by the international marketplace.

The process of degrading a tropical forest begins with a road (Figure 4-6). Once the forest becomes accessible, it is usually cut or degraded and fragmented into vulnerable patches by the following activities:

- *Unsustainable small-scale farming*. Colonists follow logging roads into the forest to plant crops on small cleared plots, to build homes, and to try to survive. With little experience in potentially sustainable slash-and-burn and shifting cultivation (Figure 3-4), many of these newcomers cut and burn too much forest to grow crops and don't allow depleted soils to recover, ultimately degrading large tracts of forest.
- *Cattle ranching*. Cattle ranches are often established on exhausted and abandoned cropland, often aided by government subsidies to make ranching profitable. Overgrazing further degrades the land.
- *Commercial logging*. Since 1950, the consumption of tropical lumber has risen 14-fold, with Japan now accounting for 60% of annual exports (followed by the United States and Great Britain). The World Bank estimates that by 2000, only 10 of the 33 countries now exporting tropical timber will have any left to export. Although timber exports to MDCs contribute to tropical forest depletion and degradation, over 80% of the trees cut in LDCs are used at home.
- *Raising cash crops*. Tropical forests are cut and converted to immense plantations used to grow crops such as sugarcane, bananas, tea, and coffee, mostly for export to MDCs.

A: Development of genetic resistance to the chemicals by target pests

- *Mining operations.* Most of the extracted minerals, such as iron ore and bauxite (aluminum ore), are exported to MDCs.
- *Oil drilling and extraction.*
- *Damming rivers and flooding large areas of forest.*

SOLUTIONS: REDUCING TROPICAL DEFORESTATION Environmentalists have suggested the following ways to reduce tropical deforestation:

- *Use remote-sensing satellites to find out how much of the world is covered with forest and how much has been deforested.* This could be done for about what the world spends for military purposes every three minutes.
- *Establish a mandatory international system for identifying tropical (and other) timber grown and harvested sustainably.* So far, only 0.1% of the world's tropical forests are managed sustainably.
- *Reform tropical timber-cutting regulations and practices.* New logging contracts would charge more for timber-cutting concessions and require companies to post adequate bonds for restoration and reforestation.
- *Fully fund the Rapid Assessment Program (RAP),* which sends biologists to assess the biodiversity of "hot spots"—forests and other habitats that are both rich in unique species and in imminent danger—with the goal of channeling funds and efforts toward immediate protection of these endangered ecosystems.
- *Use debt-for-nature swaps and conservation easements to encourage countries to protect tropical forests or other valuable natural systems.* In a debt-for-nature swap, participating tropical countries act as custodians for protected forest reserves in return for foreign aid or debt relief. With conservation easements, a country, a private organization, or a group of countries compensates individual countries for protecting selected forest areas.
- *Help settlers learn how to practice small-scale sustainable agriculture.*
- *Stop funding tree and crop plantations, ranches, roads, and destructive types of tourism on any land now covered by old-growth tropical forests.*
- *Concentrate farming, tree and crop plantations, and ranching on cleared or degraded tropical forest areas that are in various stages of secondary ecological succession* (Figure 4-3).
- *Set aside large protected areas for indigenous tribal peoples.* Indigenous peoples are the primary guardians and sustainable users of vast, mostly undisturbed habitats. These peoples are being driven from their homelands in tropical forests and other biomes by commercial resource extractors and the landless poor.
- *Pressure banks and international lending agencies* (controlled by MDCs) *not to lend money for environmentally destructive projects*—especially road building (Figure 4-6)—involving old-growth tropical forests.
- *Reduce poverty and the flow of the landless poor to tropical forests by slowing population growth.*
- *Reforest and rehabilitate degraded tropical forests and watersheds.*
- *Work with local people to protect forests* (Solutions, p. 63).

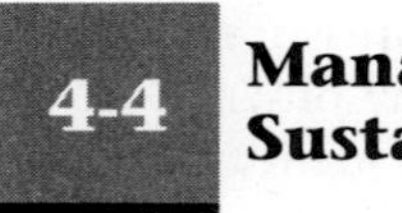

4-4 Managing and Sustaining Rangelands

THE WORLD'S RANGELAND RESOURCES Almost half of the earth's ice-free land is **rangeland**: land that supplies forage or vegetation for grazing (grass-eating) and browsing (shrub-eating) animals and that is not intensively managed. Most rangelands are grasslands in areas too dry for growing crops without irrigation.

About 42% of the world's rangeland is used for grazing livestock. Much of the rest is too dry, cold, or remote from population centers to be grazed by large numbers of livestock. About 34% of the total land area of the United States is rangeland, most of it grasslands in the arid and semiarid western half of the country.

Most rangeland grasses require rich soils (Figure 3-2) and have deep, complex root systems that not only anchor the plants but also sustain them through several seasons. Unlike most plants, the blades of rangeland grasses grow from the base, not from the leaf tip. So long as only its upper half is eaten and its lower half remains, rangeland grass can grow back quickly and is a renewable resource that can be grazed again and again.

Each type of grassland has a **carrying capacity**: the maximum number of wild or domesticated herbivores that can graze a given area without destroying the base needed for grass renewal. Carrying capacity is influenced by season, range conditions, climatic conditions, past grazing use, soil type, kinds of grazing animals, and amount of grazing.

Overgrazing occurs when too many animals graze too long and exceed the carrying capacity of a grassland area. Large populations of wild ruminants can overgraze rangeland in prolonged dry periods,

Q: What is the best way to control pests?

How Farmers and Loud Monkeys Saved a Forest

SOLUTIONS

It's early morning in a tropical forest in the Central American country of Belize. Suddenly, loud roars that trail off into wheezing moans—territorial calls of black howler monkeys—wake up everyone in or near the wildlife sanctuary by the Belize River. Vegetarians, these primates travel slowly among the treetops foraging for food.

This species is the centerpiece of an experiment integrating ecology and economics by allowing local villagers to make money by helping sustain the forest and its wildlife. The project is the brainchild of American biologist Robert Horwich. In 1985 he suggested that villagers establish a sanctuary that would benefit both the local black howlers and themselves. He suggested that the farmers leave thin strips of forest along the edges of their fields to provide food for the howlers, who feed on leaves, flowers, and fruits as they travel among the treetops.

To date, more than 100 farmers have participated, and the 47-square-kilometer (18-square-mile) sanctuary is now home for an estimated 1,100 black howlers. The idea has spread to seven other villages.

Now, as many as 6,000 ecotourists visit the sanctuary each year to catch glimpses of its loud monkeys and other wildlife. Villagers serve as tour guides, cook meals for the visitors, and lodge tourists overnight in their spare rooms.

but most overgrazing is caused by excessive numbers of domestic livestock feeding too long in a particular area. Overgrazing by livestock is the major cause of *desertification*, a serious and growing problem in many parts of the world (Figure 3-7).

SOLUTIONS: MANAGING RANGELANDS SUSTAINABLY Ranchers and nomadic herders can prevent overgrazing by controlling the *stocking rate*—the number of each particular kind of animal placed on a given area—so that it doesn't exceed carrying capacity. Determining the carrying capacity of a range site is difficult and costly, however, and carrying capacity changes because of drought, invasions by new species, and other environmental factors.

Not only the numbers, but also the distribution of grazing animals over a rangeland, must be controlled to prevent overgrazing. Ranchers can control distribution by fencing off damaged rangeland, rotating livestock from one grazing area to another, providing supplemental feeding at selected sites, and locating water holes and salt blocks in strategic places.

Another aspect of rangeland management for livestock is predator control. For decades, hundreds of thousands of predators have been shot, trapped, and poisoned in the United States by ranchers, farmers, and federal predator control officials. Many wildlife conservationists have fought unsuccessfully to have such programs eliminated because of their adverse effects on wildlife.

Some environmentalists call for a total ban on the grazing of livestock on public land, but others believe that with proper management, ranching on public and privately owned rangeland is a potentially sustainable operation. They believe that encouraging sustainable ranching practices keeps much of the land from being broken up and converted to developments. They call for curbing overgrazing on public rangeland by the following means:

- *Excluding or strictly limiting livestock grazing in riparian areas*—vulnerable strips of vegetation along bodies of water used as sources of drinking water for livestock and wildlife.
- *Banning livestock grazing on rangeland in poor condition until it recovers.*
- *Ending federal predator control measures.*
- *Greatly increasing funds for restoration of degraded rangeland and riparian areas.*
- *Sharply raising grazing fees to a fair market value.* Since 1981, grazing fees on public rangeland have been set by Congress at only one-fourth to one-eighth the going rate for leasing comparable private land. This means that taxpayers give the 2% of U.S. ranchers with federal grazing permits subsidies amounting to about $50 million a year—the difference between the fees collected and the actual value of the grazing on this land.
- *Giving family ranchers with small ranches grazing fee discounts.* This would help them stay in business and keep their land from being converted to real-estate developments. Ranchers with large operations who artificially divide their holdings to qualify would not get discounts.
- *Abolishing grazing advisory boards.* These boards give ranchers undue influence on federal officials and lead to about 96% of the allotted funds being

A: Use integrated pest management (IPM)

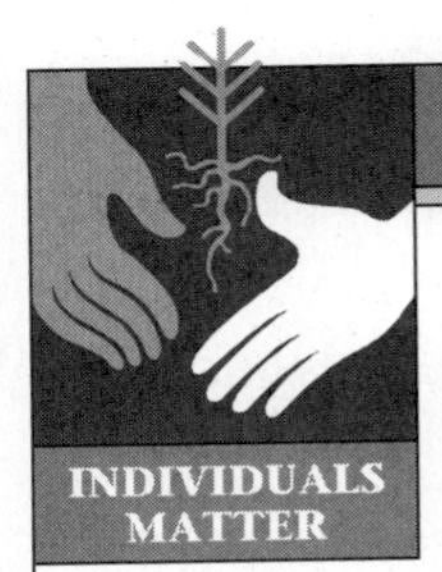

The Eco-Cowboy

INDIVIDUALS MATTER

Wyoming rancher Jack Turnell is a new breed of cowboy who talks about riparian ecology and biodiversity as fluently as he talks about cattle: "I guess I have learned how to bridge the gap between the environmentalists, the bureaucracies, and the ranching industry."

Turnell grazes cattle on his 32,000-hectare (80,000-acre) ranch south of Cody, Wyoming, and on 16,000 hectares (40,000 acres) of Forest Service land on which he has grazing rights. For the first decade after he took over the ranch he punched cows the conventional way; since then he's made some changes.

Turnell disagrees with the proposals by environmentalists to raise grazing fees and to remove sheep and cattle from public rangeland. He believes that if ranchers are kicked off the public range, ranches like his will be sold to developers and chopped up into vacation sites, irreversibly destroying the range for wildlife and livestock alike.

At the same time, he believes that ranches can be operated in more ecologically sustainable ways. To demonstrate this Turnell began systematically rotating his cows away from the riparian areas, gave up most uses of fertilizers and pesticides, and crossed his Hereford and Angus cows with a French breed that does not like to congregate around water. Most of his ranching decisions are made in consultation with range and wildlife scientists, and changes in range condition are carefully monitored with photographs.

The results have been impressive. Riparian areas on the ranch and Forest Service land are lined with willows and other plant life, providing lush habitat for an expanding population of wildlife, including pronghorn antelope, deer, moose, elk, bear, and mountain lions. And this "eco-cowboy" now makes more money because the better-quality grass puts more meat on his cattle. He frequently talks to other ranchers about sustainable range management; some of them probably think he has been chewing locoweed.

spent on grazing instead of for other uses. Other multiple-use public lands have no such boards.

- *Replacing the current noncompetitive grazing permit system with a competitive bidding system.* The current system gives ranchers that obtain essentially lifetime permits to graze livestock on public land an unfair economic advantage over ranchers who can't get such permits, which amount to government subsidies. Competitive bidding would let free enterprise work and might allow conservation and wildlife groups to obtain grazing permits they could use to protect such lands from overgrazing. Currently, it's estimated that taxpayers are providing the 30,000 ranchers holding permits with an annual subsidy of about $2 billion—an average of $67,000 per rancher—to produce only 3% of the country's beef. Critics charge that the public-lands grazing program is little more than "cowboy welfare," mostly for well-to-do ranchers who hold 90% of the permits.

The problem is that western ranchers wield enough political power to see that measures like those just listed are not enacted by elected officials. Still, some ranchers have already demonstrated that rangeland can be grazed sustainably and profitably (Individuals Matter, above).

4-5 Managing and Sustaining Parks

PARKS AROUND THE WORLD Today, over 1,100 national parks larger than 1,000 hectares (2,500 acres) each are located in more than 120 countries. Together they cover an area equal to that of Alaska, Texas, and California combined. This important achievement in global conservation was spurred by the creation of the first public national park system in the United States in 1912.

The U.S. national park system is dominated by 50 national parks, most of them in the West. These repositories of majestic beauty and biodiversity, sometimes called America's crown jewels, are supplemented by state, county, and city parks.

THE PLIGHT OF PARKS TODAY Parks everywhere are under siege. In LDCs, parks are often invaded by local people who desperately need wood, cropland, and other resources. Poachers (illegal hunters) kill animals to get and sell rhino horns, elephant tusks, furs, and other valuable parts. Park services in LDCs typically have too little money and staff to fight these invasions, either by force or by education. Also, most

Q: What percentage of the USDA's research and education budget is spent on integrated pest management?

Who's Afraid of the Big Gray Wolf?

SPOTLIGHT

At one time the gray wolf ranged over most of North America. Between 1850 and 1900, some 2 million wolves were shot, trapped, poisoned—and even drenched with gasoline and set afire—by ranchers, hunters, and government employees. The idea was to make the West and the Great Plains safe for livestock and for big-game animals prized by hunters. The gray wolf is now listed as endangered in all 48 lower states except Minnesota, whose 1,550–1,750 wolves are considered threatened.

Ecologists now recognize the important role these predators once played in parts of the West and the Great Plains. These wolves culled herds of bison, elk, and mule deer, which without predators can proliferate and devastate vegetation, which threatens the existence of other forms of wildlife.

In 1987 the U.S. Fish and Wildlife Service proposed that wolves be reintroduced into the Yellowstone ecosystem, which includes two national parks, seven national forests, and other federal and state lands in Wyoming, Idaho, and Montana. This proposal to reestablish ecological connections that humans had eliminated brought outraged howls—both from ranchers (who feared the wolves would attack their cattle and sheep) and from hunters (who feared that the wolves would kill too many big-game animals). One enraged rancher said that the idea was "like reintroducing smallpox."

An economic study estimated that returning wolves to the Yellowstone ecosystem would bring $18 million into the local economy during the first year and about $100 million over 20 years. National Park Service officials promised to trap or shoot any wolves that killed livestock outside of the recovery areas and to reimburse ranchers for lost stock from a private $100,000 fund established by Defenders of Wildlife. However, these promises fell on deaf ears, and these ranchers and hunters have worked hard to delay or defeat the plan in Congress.

Some ranchers and hunters say that they'll take care of the wolves quietly—what they call the "shoot, shovel, and shut up" solution. In 1993 a single gray wolf found in the Yellowstone ecosystem just outside the southern boundary of Yellowstone National Park was shot and killed. Do you think the gray wolf should be reintroduced to the Yellowstone ecosystem?

of the world's national parks are too small to sustain many of the larger animal species.

Popularity is the biggest problem of national and state parks in the United States (and other MDCs). Because of increased numbers of sufficiently affluent people, roads, and cars, annual recreational visits to national park system units have increased more than 12-fold (and visits to state parks seven-fold) since 1950, and the number of visits is expected to double between 1990 and 2020.

During the peak summer season, the most popular national and state parks are often choked with cars and trailers, and they are plagued by noise, traffic jams, litter, vandalism, poaching, deteriorating trails, polluted water, and crime. Many visitors to heavily used parks leave the city to commune with nature, only to find the parks noisier, more congested, and more stressful than where they came from. Park Service rangers now spend an increasing amount of their time on law enforcement instead of on resource conservation, management, and education.

Wolves (Spotlight, above), bears, and other large predators in and near various parks have all but vanished because of excessive hunting, poisoning by ranchers and federal officials, and the limited size of most parks. As a result, populations of species these predators once helped control have exploded, destroying vegetation and crowding out other species.

Alien species have moved or have been introduced into parks. Wild boars (imported to North Carolina in 1912 for hunting) are threatening vegetation in part of the Great Smoky Mountains National Park. The Brazilian pepper tree has invaded Florida's Everglades National Park. Mountain goats in Washington's Olympic National Park trample native vegetation and accelerate soil erosion.

The greatest threat to many U.S. parks today is posed by nearby human activities. Wildlife and recreational values are threatened by mining, logging, grazing, coal-burning power plants, water diversion, and urban development. Polluted air drifts hundreds of kilometers to kill trees in California's Sequoia National Park and blur the awesome vistas at Arizona's Grand Canyon. According to the National Park Service, air pollution affects scenic views in national parks more than 90% of the time. And unless a massive and expensive ecological restoration project works, Florida's Everglades National Park may dry up

A: 1%

Biocultural Restoration in Costa Rica

SOLUTIONS

In a rugged mountainous region covered with tropical rain forest lies a centerpiece of Costa Rica's efforts to preserve its biodiversity—the Guanacaste National Park, which has been designated an international biosphere reserve. In the park's lowlands a small tropical seasonal forest is being restored and relinked to the rain forest on adjacent mountain slopes.

Daniel Janzen, professor of biology at the University of Pennsylvania, has helped galvanize international support and has raised more than $10 million for this restoration project, the world's largest. Janzen is a leader in the growing field of rehabilitation and restoration of degraded ecosystems.

Janzen's vision is to make the nearly 40,000 people who live near the park an essential part of the restoration of the degraded forest—a concept he calls *biocultural restoration*. By actively participating in the project, local residents reap enormous educational, economic, and environmental benefits. Local farmers have been hired to sow large areas with tree seeds and to plant seedlings started in Janzen's lab.

Students in grade schools, high schools, and universities study the park's ecology in the classroom and then go on annual field trips to the park itself. There are educational programs for civic groups and tourists from Costa Rica and elsewhere. These visitors and their activities will stimulate the local economy. The project will also serve as a training ground in tropical forest restoration for scientists from all over the world.

Janzen recognizes that in 20 to 40 years today's children will be running the park and the local political system. If they understand the importance of their local environment, they are more likely to protect and sustain its biological resources. He understands that education, awareness, and involvement—not guards and fences—are the best way to protect ecosystems from unsustainable use.

because water is being siphoned off to grow crops and support Florida's rapidly growing human population.

SOLUTIONS: BETTER PARK MANAGEMENT IN LDCS Some park managers, especially in LDCs, are developing integrated management plans that combine conservation practices with sustainable development of resources in the parks and surrounding areas. In such plans the inner core and especially vulnerable areas of each park are to be protected from development and treated as wilderness. Restricted numbers of people are allowed to use these areas only for hiking, nature study, ecological research, and other nondestructive recreational and educational activities. In buffer areas surrounding the core, controlled commercial logging, sustainable grazing by livestock, and sustainable hunting and fishing by local people are allowed. Money spent by park visitors adds to local income. By involving local people in developing park management plans, managers seek to help them see the park as a vital resource they need to protect and sustain rather than ruin (Solutions, above).

Integrated park management plans look good on paper, but they cannot be carried out without adequate funding and the support of nearby landowners and users. Moreover, in some cases the protected inner core may be too small to sustain some of the park's larger animal species.

SOLUTIONS: AN AGENDA FOR U.S. NATIONAL PARKS In the United States, national parks are managed under the principle of natural regulation—as if they are wilderness ecosystems that, if left alone, will sustain themselves. Many ecologists consider this a misguided policy. Most parks are far too small to even come close to sustaining themselves, and even the biggest ones cannot be isolated from the harmful effects caused by activities in nearby areas.

In 1988 the Wilderness Society and the National Parks and Conservation Association suggested the following blueprint for the future of the U.S. national park system:

- *Educate the public about the urgent need to protect, mend, and expand the system.*
- *Significantly increase the number and pay of park rangers.*
- *Acquire new parkland near threatened areas and add at least 75 new parks within the next decade.*
- *Locate most commercial park facilities (such as restaurants, hotels, and shops) outside park boundaries.*
- *Raise the fees charged to private concessionaires who operate lodging, food, and recreation services inside national parks to at least 22% of their gross receipts.* The current maximum return averages only 2.5% of the $1.5 billion concessionaires take in annual-

Q: What percentage of the earth's land area is covered by forests?

ly. Many large concessionaires have long-term contracts by which they pay the government as little as 0.75% of their gross receipts.

- *Halt concessionaire ownership of facilities in national parks, which makes buying buildings back very expensive.*
- *Wherever feasible, place visitor parking areas outside the park boundary.* Use low-polluting vehicles to transport visitors to and from parking areas and within the park.
- *Greatly expand the National Park Service budget for maintenance and for science and conservation programs.* The national parks face a $2.3-billion backlog of repairs.
- *Require the Park Service, Forest Service, and Bureau of Land Management to develop integrated management plans so that activities in nearby national forests and rangelands don't degrade national parklands and wilderness areas within the parks.*

4-6 Preserving Wilderness

WHY PRESERVE WILDERNESS? Protecting undeveloped lands or wilderness from exploitation is an important way of preserving ecological diversity. According to the U.S. Wilderness Act of 1964, **wilderness** consists of those areas "where the earth and its community of life are untrammeled by man, where man himself is a visitor who does not remain."

The Wilderness Society estimates that a wilderness area should contain at least 400,000 hectares (1 million acres); otherwise it can be affected by air, water, and noise pollution from nearby mining, oil and natural gas drilling, timber cutting, industry, and urban development. Environmentalists urge that the world's remaining wilderness be protected by law everywhere, focusing first on the most endangered spots in the most wilderness-rich and biodiverse countries (Figure 4-1).

In addition to providing recreation for growing numbers of people, wilderness also has important ecological values. It provides undisturbed habitats for wild plants and animals, protects diverse biomes from damage, and provides a laboratory in which we can discover more about how nature works. Wilderness is a biodiversity bank and an eco-insurance policy. In the words of Henry David Thoreau, "In wildness is the preservation of the world."

SOLUTIONS: THE NATIONAL WILDERNESS PRESERVATION SYSTEM In the United States, preservationists have been trying to keep wild areas from being developed since 1900. On the whole they have fought a losing battle. Not until 1964 did Congress pass the Wilderness Act, which allowed the government to protect undeveloped tracts of public land from development as part of the National Wilderness Preservation System.

Only 4% of U.S. land area is protected as wilderness, with almost three-fourths of it in Alaska. Only 1.8% of the land area of the lower 48 states is protected, most of it in the West. Of the 413 wilderness areas in the lower 48 states, only four are larger than 400,000 hectares. Furthermore, the present wilderness preservation system includes only 81 of the country's 233 distinct ecosystems. Like the national parks, most wilderness areas in the lower 48 states are vulnerable islands of habitat in a sea of development.

There remain almost 40 million hectares (100 million acres) of public lands that could qualify for designation as wilderness. Some environmentalists want all of this land protected as wilderness and advocate vigorous efforts to rehabilitate other lands to enlarge existing wilderness areas. Such efforts are strongly opposed by timber, mining, ranching, energy, and other interests who want either to extract resources from these lands or convert such public lands to private ownership.

4-7 The Importance and Plight of Wild Species

THE RISE AND FALL OF SPECIES We don't have the foggiest notion how many species of plants, animals, and microorganisms exist on Earth; estimates range from 5 million to 100 million, most of them insects, microscopic organisms, and tiny sea creatures. So far biologists have identified and named only about 1.4 million species. They know a fair amount about roughly one-third of these species and the detailed roles and interactions of only a few.

As the planet's surface and climate have changed over the 4.6 billion years of its existence, species have disappeared and new ones have evolved to take their places. Biologists estimate that 99% of all the species that have ever lived are now extinct.

Some species inevitably disappear as local conditions change, a process called *background extinction*. In contrast, *mass extinction* is an abrupt rise in extinction rates above the background level. It is a catastrophic,

The Passenger Pigeon: Gone Forever

CASE STUDY

In the early 1800s, bird expert Alexander Wilson watched a single migrating flock of passenger pigeons darken the sky for over four hours. He estimated that this flock was more than 2 billion birds strong, 386 kilometers (240 miles) long, and 1.6 kilometers (1 mile) wide.

By 1914 the passenger pigeon (Figure 4-7) had disappeared forever. How could the species that was once the most common bird in North America become extinct within 100 years?

The answer is humans. The main reasons for the extinction of this species were uncontrolled commercial hunting and loss of the bird's habitat and food supply as forests were cleared for farms and cities.

Passenger pigeons were good to eat, their feathers made good pillows, and their bones were widely used for fertilizer. They were easy to kill because they flew in gigantic flocks and nested in long, narrow colonies.

Beginning in 1858, passenger pigeon hunting became a big business. Shotguns, traps, artillery, and even dynamite were used. Birds were suffocated by burning grass or sulfur below their roosts. Live birds were used as targets in shooting galleries. In 1878 one professional pigeon trapper made $60,000 by killing 3 million birds at their nesting grounds near Petoskey, Michigan.

By the early 1880s commercial hunting had ceased because only a few thousand birds were left. At that point recovery of the species was doomed because females laid only one egg per nest. On March 24, 1900, a young boy in Ohio shot the last known wild passenger pigeon. The last passenger pigeon on earth, a hen named Martha after Martha Washington, died in the Cincinnati Zoo in 1914. Her stuffed body is now on view at the National Museum of Natural History in Washington, D.C.

widespread (often global) event in which not just one species but large groups of species are wiped out. Fossil and geological evidence indicates that the earth's species have experienced five great mass extinctions—at roughly 26-million-year intervals—with smaller ones in between.

Evidence also indicates that these mass extinctions have been followed by periods of recovery and *adaptive radiations,* in which numerous new species evolved to fill new or vacant ecological niches in changed environments. Speciation minus extinction equals *biodiversity,* one of the planet's most important resources.

WHY PRESERVE WILD SPECIES? Millions of species have vanished over the earth's long history, so why should we worry about losing a few more? Does it matter that the passenger pigeon (Case Study, above), the great auk, or some unknown plant or insect in a tropical forest has become extinct, mostly because of human activities (Figure 4-7)? Does it matter that the existences of the California condor (only seven birds in the wild), the whooping crane (142 in the wild), and hundreds of other species are threatened because of human activities (Figure 4-8, pp. 70–71)?

The answer in both instances is yes. Here are several reasons for protecting wild species from premature extinction as a result of human activities, thereby allowing them to play their roles in the ongoing saga of evolution:

- *Economic importance.* About 10% of the U.S. gross domestic product comes directly from use of wild species. Some 90% of today's food crops were domesticated from wild tropical plants. Agricultural scientists and genetic engineers will need to use existing wild plant species, most of them still unknown, as sources of food and to develop new crop strains. Wild plants and plants domesticated from wild species supply rubber, oils, dyes, fiber, paper, lumber, and other useful products. Nitrogen-fixing microbes in the soil and in plants' root nodules supply nitrogen to grow food crops worth almost $50 billion per year worldwide ($7 billion in the United States). Pollination by birds and insects is essential to many food crops, including 40 U.S. crops valued at approximately $30 billion per year.
- *Medical importance.* About 75% of the world's population relies on plants or plant extracts for medicines. Roughly half of the medicines used in the world, and 25% of those used in the United States, have active ingredients extracted from wild species. Many animal species are used to test drugs, vaccines, chemical toxicity, and surgical procedures, and in studies of human health and disease. Under intense pressure from animal rights groups, scientists are seeking alternate testing methods that minimize animal suffering or, better yet, do not use animals at all. However,

Q: How much of all U.S. land consists of public lands?

Passenger pigeon

Great auk

Dodo

Bushy seaside sparrow

Aepyornis (Madagascar)

Figure 4-7 Some species that have become extinct largely because of human activities, mostly habitat loss and overhunting.

they caution that alternative techniques cannot replace all animal research.

- *Aesthetic and recreational importance.* Wild plants and animals are a source of beauty, wonder, joy, and recreational pleasure for many people. Wildlife tourism, sometimes called *ecotourism*, generates as much as $12 billion in revenues each year. One wildlife economist has estimated that one male lion living to age 7 generates $515,000 in tourism in Kenya; by contrast, if killed for its skin, the lion would bring only about $1,000. However, care must be taken to ensure that ecotourists neither damage or disturb wildlife and ecosystems nor disrupt local cultures.
- *Scientific importance.* Each species has scientific value because it can help scientists understand how life has evolved—and how it will continue to evolve—on this planet.
- *Ecological importance.* Wild species supply us (and other species) food from the soil and the sea, recycle nutrients essential to agriculture, and help maintain soil fertility. They also produce oxygen and other gases in the atmosphere, moderate the earth's climate, help regulate water supplies, and store solar energy. Moreover, they detoxify poisonous substances, break down organic wastes, control potential crop pests and disease carriers, and make up a vast gene pool from which we and other species can draw.
- *Ethics.* To some people, each wild species has an inherent right to exist—or at least to struggle to exist. According to this view, it is morally wrong for us to hasten the extinction of any species. Some people distinguish between the survival rights of plants and those of animals, mostly for what they deem practical reasons. Poet Alan Watts, for example, once said that he was a vegetarian "because cows scream louder than carrots."

THE CURRENT EXTINCTION CRISIS: FATAL SUBTRACTION Imagine that you are driving down a highway at a high speed and that all the while your passengers are busily dismantling parts of your car at random, and throwing them out the window. How long will it be before they remove enough parts to cause a crash?

This urgent question, applied to the earth, is one that we as a species should be asking ourselves. As we tinker with the only home for us and other species, we are rapidly removing parts of the earth's natural biodiversity, upon which we and other species depend in ways we know little about. We are not heeding Aldo Leopold's warning: "To keep every cog and wheel is the first precaution of intelligent tinkering."

Sooner or later all species become extinct, but humans have become a primary factor in the premature extinction of more and more species as we march relentlessly across the globe (Figure 4-7). It's difficult to document extinctions, for most go unrecorded. But biologists estimate that during 1993 at least 4,000 and as many as 36,000 species became extinct (an average of 11 to 100 species per day); the figure could reach 50,000 species per year by 2000. These scientists warn that if deforestation (especially of tropical forests), desertification, and destruction of wetlands and coral reefs continue at their current rates, within the next few decades we could easily lose at least a quarter, and conceivably half, of the earth's species.

Mass extinctions occurred long before we arrived, but there are two important differences between the present mass extinction and those of the past: **(1)** *the present extinction crisis is the first to be caused by a single species—our own*, and **(2)** *the current wildlife holocaust is taking place in only a few decades, rather than over thousands to millions of years.*

ENDANGERED AND THREATENED SPECIES

Species heading toward extinction are classified as either endangered or threatened (Figure 4-8). An

A: 42% (35% managed by the federal government)

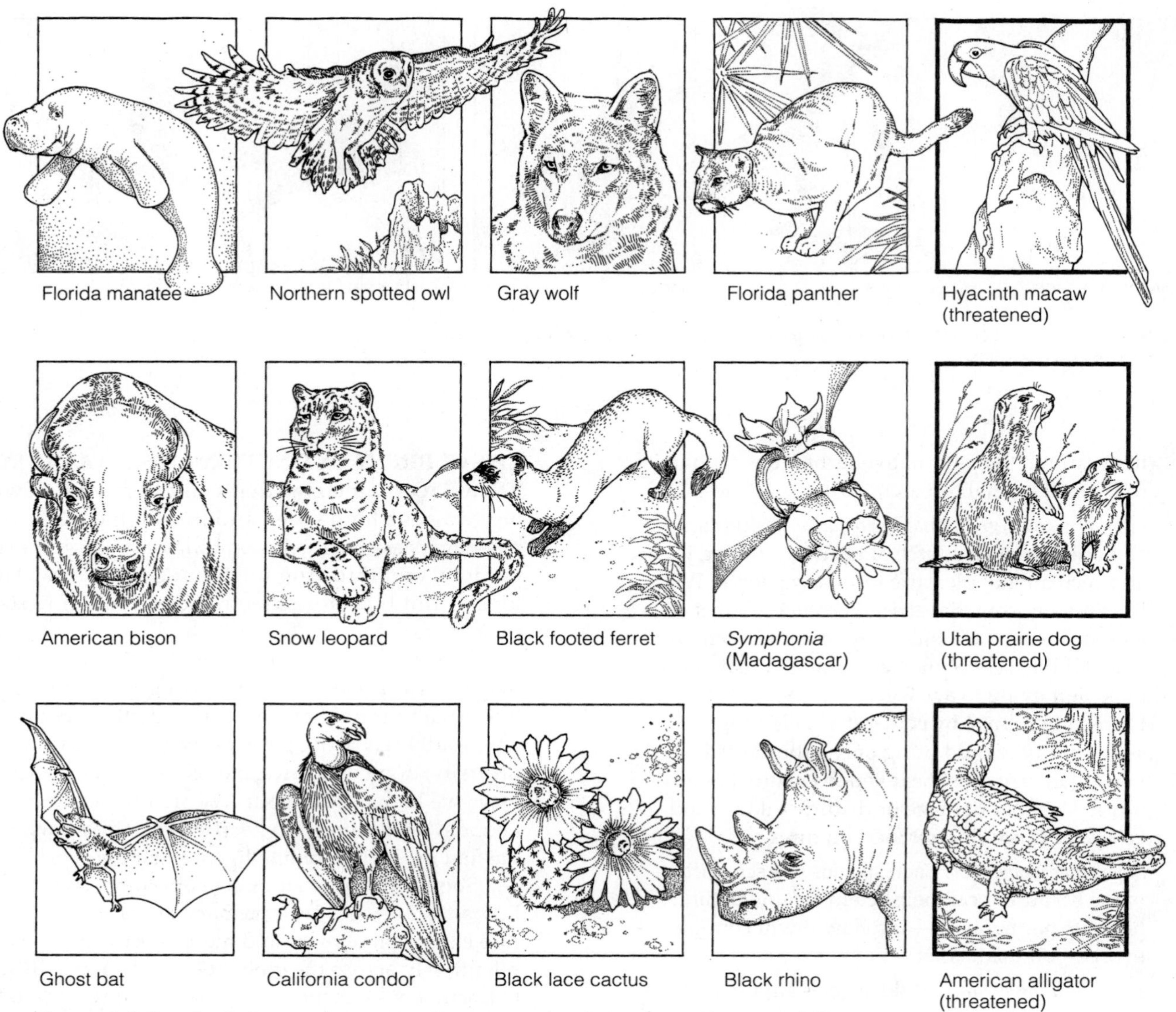

Figure 4-8 Species that are endangered or threatened primarily because of human activities, mostly habitat loss and overhunting.

endangered species has so few individual survivors that the species could soon become extinct over all or most of its natural range. A **threatened species** is still abundant in its natural range but is declining in numbers and likely to become endangered.

Some species have one or more characteristics that make them more vulnerable than others to premature extinction (Connections, p. 73). Examples of these characteristics are **(1)** *low reproductive rate* (blue whale, California condor, giant panda, whooping crane; Figure 4-8), **(2)** *specialized feeding habits* (blue whale, black-footed ferret, giant panda, Australian koala, bats), **(3)** *large size* (African Bengal tiger, rhinoceros, giant panda, grizzly bear), **(4)** *limited or specialized nesting or breeding areas* (whooping crane, green sea turtles, bald eagle, Kirtland's warbler, cave-dwelling bats), and **(5)** *fixed migratory patterns* (blue whale, whooping crane, Kirtland's warbler).

The populations of many wild species, while not yet in danger of extinction, have diminished locally or regionally. Such species may be a better indicator of the condition of entire ecosystems than are endangered and threatened species. They can serve as early warnings so that we can prevent extinctions rather than responding to emergencies.

MAJOR CAUSES OF PREMATURE WILDLIFE LOSS

The major causes of extinction and population reduction of wildlife include the following:

- *Human population growth* (Figure 1-1).

Q: Where is most of the federally owned and managed land in the United States?

- *Increased resource use* (Figure 1-2).
- *Economic systems and policies that fail to value the environment and its vital ecosystem services*—and thus promote unsustainable exploitation (Solutions, p. 5).
- *Habitat loss and fragmentation* (Figure 4-9). Deforestation (Figures 4-4 and 4-5) is the greatest eliminator of species, followed by destruction of coral reefs and wetlands and plowing of grasslands. Furthermore, much of the remaining wildlife habitat is being fragmented into patches, or "habitat islands," that may be too small to support the minimum breeding populations of species.
- *Commercial hunting and poaching*. Bengal tigers are in trouble because a tiger fur sells for $100,000 in Tokyo. A mountain gorilla is worth $150,000; an ocelot skin $40,000; an Imperial Amazon macaw $30,000; a snow leopard skin $14,000. Rhinoceros horn sells for as much as $28,600 per kilogram. As more species become endangered, the demand for them on the black market soars, hastening their extinction. Poaching of endangered or threatened species (mostly for Asian markets) is increasing in the United States, especially in western national parks and wilderness areas covered by only 22 federal wildlife protection officers. Most poachers are not caught, and the money to be made far outweighs the risk of fines and jail.

A: Alaska (73%) and the western states (22%)

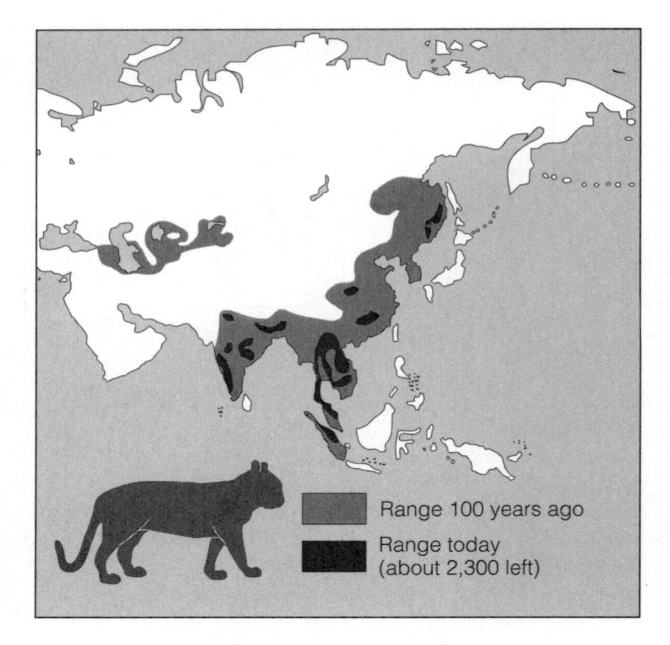

Indian Tiger

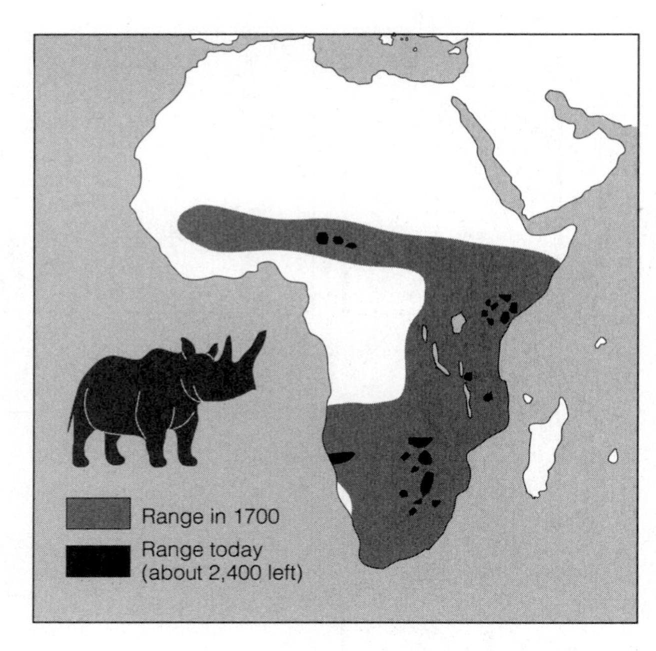

Black Rhino

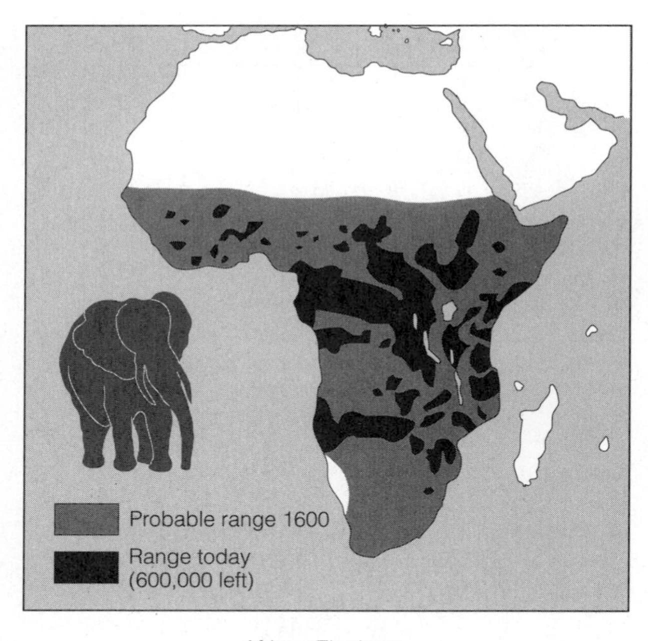

African Elephant

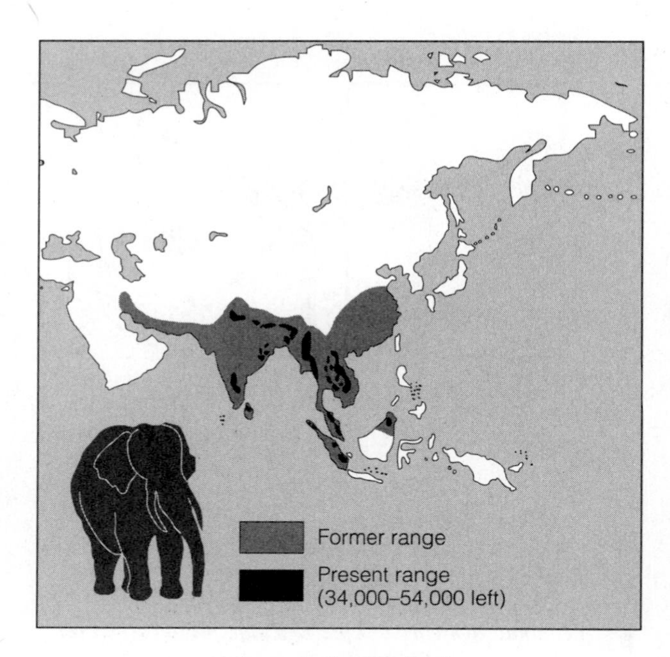

Asian or Indian Elephant

Figure 4-9 Reduction in the range of several species, mostly through a combination of habitat loss and hunting. What will happen to these and millions of other species when the global human population doubles in the next few decades? (Data from International Union for the Conservation of Nature and World Wildlife Fund)

- *Predator and pest control.* People try to exterminate species that compete with us for food and game. African farmers kill large numbers of elephants to keep them from trampling and eating food crops. Ranchers, farmers, and hunters in the United States support control of species such as coyotes and wolves (Spotlight, p. 65) that can prey on livestock and game species prized by hunters. Since 1929, U.S. ranchers and government agencies have poisoned 99% of North America's

Q: What percentage of the earth's land area is covered by tropical forests?

CONNECTIONS

Bats: A Bad Rap

Despite their variety (950 species) and worldwide distribution, bats have several traits that expose them to extinction because of human activities. Many bats nest in huge cave colonies, which become vulnerable to destruction when people block the cave's entrances. And once the population of a particular bat species falls below a certain level, it may not recover because bats have a low reproductive rate.

Bats play significant ecological roles and are also of great economic importance. Some bat species help control many crop-damaging insects and other pest species, such as mosquitoes. About 70% of all bat species feed on night-flying insects, making them the primary controls for such insects.

Other species of bats eat pollen; still others eat certain fruits. Bats with this kind of specialized feeding are the chief pollinators for certain trees, shrubs, and other plants; they also spread plants throughout tropical forests by excreting undigested seeds. If these keystone species are eliminated from an area, dependent plants would disappear. If you enjoy bananas, cashews, dates, figs, avocados, or mangos, you can thank bats. Research on bats has contributed to the development of birth control and artificial insemination methods, the testing of drugs, studies of disease resistance and aging, vaccine production, and development of navigational aids for the blind.

People mistakenly fear bats as filthy, aggressive, rabies-carrying blood-suckers. But most bat species are harmless to people, livestock, and crops. Only three species of bats (none of them found in the United States) feed on blood, mostly that of cattle or wild animals. These bat species can be serious pests to domestic livestock but rarely affect humans.

Because of unwarranted fears of bats and misunderstanding of their vital ecological roles, several species have been driven to extinction, and others, such as the ghost bat (Figure 4-8), are endangered or threatened. We need to see bats as valuable allies, not as enemies.

prairie dogs (Figure 4-8) because horses and cattle sometimes step into the burrows and break their legs. This has also nearly wiped out the endangered black-footed ferret (Figure 4-8), which preyed on the prairie dog.

- *The market for pets and decorative plants.* Worldwide, over 5 million live wild birds are captured and sold legally each year, and 2.5 million more are captured and sold illegally. Over 40 species, mostly parrots, are endangered or threatened because of this wild-bird trade. For every wild bird that reaches a pet shop legally or illegally, as many as 10 others die during capture or transport. Some exotic plants, especially orchids and cacti (such as the black lace cactus, Figure 4-8), are endangered because they are gathered, often illegally, sold at high prices to collectors, and used to decorate houses, offices, and landscapes.
- *Climate change and pollution.* A potential problem for many species is the possibility of fairly rapid (40–50 year) changes in climate accelerated by deforestation and emissions of heat-trapping gases into the atmosphere (Section 6-2). Another threat is toxic chemicals that can kill or harm some plants and animals (Figure 3-10).
- *Introduced species.* Exotic species are often accidentally or intentionally introduced to new geographical regions. Many of these introduced species provide food and aesthetic beauty, and they help control pests in their new environments. Some exotic species, however, have no natural predators and competitors in their new habitats, which allows them to dominate their new ecosystem and reduce or eliminate the populations of many native species (Case Study, p. 74).

4-8 Solutions: Protecting Wild Species from Extinction

TREATIES AND LAWS Several international treaties and conventions help protect endangered or threatened wild species. One of the most far-reaching is the 1975 Convention on International Trade in Endangered Species (CITES). This treaty, now signed by 119 countries, lists 675 species that cannot be commercially traded as live specimens or wildlife products because they are endangered or threatened.

Unfortunately, enforcement of this treaty is spotty, convicted violators often pay only small fines, and

The Water Hyacinth

CASE STUDY

One example of the effects of the accidental introduction of an exotic species is the story of the fast-growing water hyacinth, which is native to Central and South America. In 1884 a woman took a plant from a New Orleans exhibition and planted it in her backyard in Florida. Unchecked by natural enemies and thriving on Florida's nutrient-rich waters, within 10 years water hyacinth plants (which can double their population every two weeks) rapidly displaced native plants, clogging many ponds, streams, canals, and rivers—first in Florida and later elsewhere in the southeastern United States.

Mechanical harvesters and herbicides have failed to keep the plant in check. Although grazing Florida manatees (sea cows) can control water hyacinths better than mechanical or chemical methods, these gentle and playful herbivores are threatened with extinction (Figure 4-8). Slashed by powerboat propellers, entangled in fishing gear—even hit on the head by oars—they reproduce too slowly to recover from these assaults and loss of habitat.

Scientists have introduced water hyacinth-eaters to help control its spread. The control agents include a weevil from Argentina, a water snail from Puerto Rico, and the grass carp, a fish from the former Soviet Union. These species can help, but water snails and grass carp also feed on other, desirable aquatic plants.

The good news is that water hyacinths provide several benefits. They absorb toxic chemicals in sewage treatment lagoons. They can be fermented into a fuel similar to natural gas, added as a mineral and protein supplement to cattle feed, and applied to the soil as fertilizer. They can also be used to clean up polluted ponds and lakes—if their growth can be kept under control.

member countries can exempt themselves from protection of any listed species. Also, much of the $5-billion-per-year illegal trade in wildlife and wildlife products goes on in countries such as Singapore that have not signed the treaty. Other centers of illegal animal trade are Argentina, Indonesia, Spain, Taiwan, and Thailand.

The United States controls imports and exports of endangered wildlife and wildlife products with two important laws. The Lacey Act of 1900 prohibits transporting live or dead wild animals or their parts across state borders without a federal permit.

The Endangered Species Act of 1973 (amended in 1982 and 1988) makes it illegal for U.S. citizens to import or trade in any product made from an endangered or threatened species unless it is used for an approved scientific purpose or to enhance the survival of the species. It also authorizes the National Marine Fisheries Service (NMFS) to identify and list endangered and threatened ocean species; the Fish and Wildlife Service (FWS) identifies and lists all other endangered and threatened species. These species cannot be hunted, killed, collected, or injured in the United States.

Any decision by either agency to list or unlist a species must be based on biology only, not on economic considerations. The Endangered Species Act also forbids federal agencies to carry out, fund, or authorize projects that would either jeopardize an endangered or threatened species or destroy or modify its critical habitat—the land, air, and water necessary for its survival.

Between 1970 and 1992 the number of species found only in the United States that have been placed on the official endangered and threatened list increased from 92 to 750. Also on the list are 529 species found elsewhere.

Getting listed is only half the battle. Next the FWS or the NMFS is supposed to prepare a plan to help the species recover. However, because of a lack of funds, recovery plans have been developed and approved for only about 61% of the endangered or threatened U.S. species, and half of those plans exist only on paper.

The annual federal budget for endangered species is less than what beer companies spend on two 30-second TV commercials during the Super Bowl. At this level of funding it will take up to 48 years to evaluate the almost 3,500 species now proposed for listing. Wildlife experts estimate that at least 400 of them will vanish while they wait, as did 34 species awaiting listing between 1980 and 1990.

The act also requires that all commercial shipments of wildlife and wildlife products enter or leave the country through one of nine designated ports. Many illegal shipments slip by, however, because the 60 FWS inspectors can physically examine only about one-fourth of the 90,000 shipments that enter and leave the United States each year. Even if caught, many violators are not prosecuted, and convicted violators often pay only a small fine.

Q: How much of the earth's tropical forests have been cleared or damaged?

Since this act was passed, intense pressure has been exerted by developers, logging and mining companies, and other users of land resources to allow consideration of economic factors, both in evaluating species for listing and also in carrying out federally funded projects that threaten the critical habitats of endangered or threatened species. Despite widespread publicity about species' roles in blocking development and resource extraction between 1987 and 1992, only 18 of more than 2,000 projects evaluated by the FWS were blocked or withdrawn as a result of the Endangered Species Act.

Environmentalists and many members of Congress, however, argue that the Endangered Species Act should be strengthened—not weakened—by the following means:

- *Emphasizing protection of entire ecosystems* to help prevent future declines in species not yet listed as threatened or endangered.
- *Setting deadlines for development and implementation of recovery plans.*
- *Giving private landowners tax write-offs or other incentives for assisting in species recovery.*
- *Greatly increasing annual funding for endangered and threatened species* from the current $55 million to $460 million—the amount needed to get the job done. This amounts to spending less than $2 per year per U.S. citizen to help protect the entire nation's biological resources.
- *Allowing citizens to file lawsuits immediately if an endangered species faces serious harm or extinction.*

Such proposals are vigorously opposed by those wishing to have greater access to resources on public and other lands.

WILDLIFE REFUGES Since 1903, when President Theodore Roosevelt established the first federal wildlife refuge at Pelican Island, Florida, the national wildlife refuge system has grown to 503 refuges. (About 85% of the area included in these refuges is in Alaska.)

Over three-fourths of the refuges are wetlands for protection of migratory waterfowl. Most species on the U.S. endangered list have habitats in the refuge system, and some refuges have been set aside for specific endangered species.

Congress has not established guidelines for management of the national wildlife refuge system, leaving it largely up to the director of the U.S. Fish and Wildlife Service. As a result, many refuges are used for hunting, fishing, trapping, timber cutting, grazing, farming, oil and gas development (Case Study, p. 76), mining, military air exercises, power and air boating, and off-road vehicles. A 1990 report by the General Accounting Office found that activities considered harmful to wildlife occur in nearly 60% of the nation's wildlife refuges.

Private groups play an important role in conserving wildlife in refuges and other protected areas. For example, since 1951 the Nature Conservancy has preserved over 1 million hectares (2.5 million acres) of forests, marshes, prairies, islands, and other areas of unique ecological or aesthetic significance in the United States.

GENE BANKS, BOTANICAL GARDENS, AND ZOOS Botanists preserve genetic information and endangered plant species by storing their seeds in gene banks—refrigerated, low-humidity environments. Scientists urge that many more such banks be established, especially in LDCs; however, some species can't be preserved in gene banks, and maintaining the banks is very expensive.

The world's 1,500 botanical gardens and arboreta hold about 90,000 plant species. However, these sanctuaries have too little storage capacity and too little funding to preserve most of the world's rare and threatened plants.

Worldwide, 500 zoos house about 540,000 individual animals, many of them from species not threatened or endangered. Zoos and animal research centers increasingly are being used to preserve some or all of the remaining individuals of critically endangered animal species, with the long-term goal of reintroducing the species into protected wild habitats. Such efforts have helped save a number of nearly extinct species (Figure 4-8), including the Arabian oryx, the California condor, the peregrine falcon, the black-footed ferret, and the golden lion tamarin.

However, maintenance of populations of endangered animal species in zoos and research centers is limited by lack of space and money. Because of shortages of funds and trained personnel, only a few of the world's endangered and threatened species can be saved by treaties, laws, wildlife refuges, and zoos, and thus wildlife experts must decide which species out of thousands of candidates should be saved. Many wildlife experts suggest that the limited funds for preserving threatened and endangered species be concentrated on those that **(1)** have the best chance for survival, **(2)** have the most ecological value to an ecosystem, and **(3)** are potentially useful for agriculture, medicine, or industry.

SOLUTIONS: INDIVIDUAL ACTION We are all involved, at least indirectly, in the destruction of wildlife or the degradation of wildlife habitats any time we buy or drive a car, build a house, consume almost anything, and waste electricity, paper, water,

Should We Develop Oil and Gas in the Arctic National Wildlife Refuge?

CASE STUDY

The Arctic National Wildlife Refuge on Alaska's North Slope (Figure 4-10), which contains more than one-fifth of all the land in the U.S. wildlife refuge system, has been called the crown jewel of the refuge system. During all or part of the year it is home to more than 160 animal species, including caribou, musk oxen, snowy owls, threatened grizzly bears (Figure 4-8), arctic foxes, and migratory birds (including as many as 300,000 snow geese). It is also home for about 7,000 Inuit (Eskimos) who depend on the caribou for a large part of their diet.

The refuge's coastal plain, its most biologically productive part, is the only stretch of Alaska's arctic coastline not open to oil and gas development. U.S. oil companies hope to change this because they believe that the area *might* contain oil and natural gas deposits. Since 1985 they have been urging Congress to open to drilling some 607,000 hectares (1.5 million acres) along the coastal plain—roughly two-thirds the size of Yellowstone National Park. They argue that such exploration is needed to provide the United States with more oil and natural gas and to reduce dependence on oil imports.

Environmentalists oppose this proposal and want Congress to designate the entire coastal plain as protected wilderness. They cite Interior Department estimates that there is only a 19% chance of finding in the coastal plain as much oil as the United States consumes every six months. Even if the oil does exist, environmentalists do not believe the potential degradation of any portion of this irreplaceable wilderness area would be worth it, especially considering that improvements in energy efficiency would save far more oil at a much lower cost (Section 5-2).

Oil company officials claim they have developed Alaska's Prudhoe Bay oil fields without significant harm to wildlife; they also contend that the area they want to open to oil and gas development is less than 1.5% of the entire coastal plain region—equivalent to an oil field the size of Dulles International Airport in Washington, D.C., within an area approximately the size of South Carolina.

However, according to a U.S. Fish and Wildlife Service study, oil development in the coastal plain could cause the loss of 20–40% of the area's 180,000-head caribou herd, 25–50% of the remaining musk oxen, 50% or more of the wolverines, and 50% of the snow geese that live there part of the year. A 1988 EPA study also found that "violations of state and federal environmental regulations and laws are occurring at an unacceptable rate" in the Prudhoe Bay area, where oil fields and facilities have been developed.

Do you think this wildlife refuge should be explored and developed for oil?

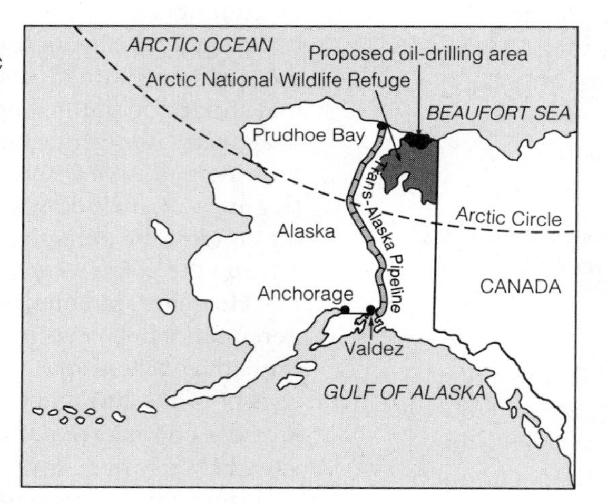

Figure 4-10 Proposed oil-drilling area in Alaska's Arctic National Wildlife Refuge. (Data from U.S. Fish and Wildlife Service)

or any other resource. Each of us can play a role in helping protect biodiversity by reducing our impact on wildlife (Individuals Matter, p. 77).

During our short time on this planet we have gained immense power over what species—including our own—live or die. We named ourselves the wise (*sapiens*) species. In the next few decades we will learn whether we are indeed a wise species—whether we have the wisdom to learn from and work with nature to protect ourselves and other species.

We abuse land because we regard it as a commodity belonging to us. When we see land as a community to which we belong, we may begin to use it with love and respect.

ALDO LEOPOLD

Q: At current loss rates, when will most remaining tropical forests be gone?

What *You* Can Do to Help Preserve Biodiversity

INDIVIDUALS MATTER

- *Improve the habitat on a patch of the earth in your immediate environment, emphasizing the promotion of biological diversity.*
- *Refuse to buy furs, ivory products, reptile-skin goods, tortoiseshell jewelry, rare orchids or cacti, and materials from endangered or threatened animal species.*
- *Leave wild animals in the wild.*
- *Reduce habitat destruction and degradation by recycling paper, cans, plastics, and other household items. Better yet, reuse items and sharply reduce your use of throwaway items.*
- *Support efforts to sharply reduce the destruction and degradation of tropical forests and old-growth forests, to slow projected global warming* (Section 6-2), *and to reduce ozone depletion in the stratosphere* (Section 6-3).
- *Pressure elected officials to pass laws requiring larger fines and longer prison sentences for wildlife poachers and to provide more funds and personnel for wildlife protection.*
- *Pressure Congress to pass a national biodiversity act and to develop a national conservation program as part of the World Conservation Strategy.*

CRITICAL THINKING

1. a. Explain why you agree or disagree with each of the proposals on p. 59 concerning protection of publicly owned timber resources in the United States.
   b. Explain why you agree or disagree with each of the proposals on p. 62 concerning protection of the world's tropical forests.
   c. Explain why you agree or disagree with each of the proposals on pp. 63–64 for use of public rangeland in the United States.
   d. Should trail bikes, dune buggies, and other off-road vehicles be banned from all public rangeland to reduce damage to vegetation and soil? Explain. Should such a ban also include national forests, national wildlife refuges, and national parks? Explain.
   e. Explain why you agree or disagree with each of the proposals listed on pp. 66–67 concerning the U.S. national park system.
2. Should more wilderness areas be preserved in the United States, especially in the lower 48 states? Explain.
3. Discuss your gut-level reaction to the following statement: "It doesn't really matter that the passenger pigeon is extinct and that the blue whale, the whooping crane, the California condor, the rhinoceros, and the grizzly bear are endangered mostly because of human activities." Be honest about your reactions and give the basis for each of them.
4. Make a log of your own consumption of products for a single day. Relate your consumption to the increased destruction of wildlife and wildlife habitats in the United States, in tropical forests, and in aquatic ecosystems.
5. Do you accept the ethical position that each *species* has the inherent right to survive without human interference, regardless of whether it serves any useful purpose for humans? Explain.

A: 30–50 years

5 Energy

If the United States wants to save a lot of oil and money and increase national security, there are two simple ways to do it: stop driving Petropigs and stop living in energy sieves.

AMORY LOVINS

5-1 The Nature of Energy Resources

ENERGY RESOURCE USE TODAY Energy is the thread sustaining and integrating all life and supporting all economies. Energy comes in many forms: light; heat; electricity; chemical energy stored in the chemical bonds in coal, sugar, wood, and other materials; moving matter such as water, wind (air masses), and joggers; and nuclear energy emitted when the tiny nuclei that make up the center of certain atoms are split apart (nuclear fission) or fused together (nuclear fusion).

Some 99% of the energy used to heat the earth, and all our buildings, comes directly from the sun (Figure 1-4). Without this direct input of renewable solar energy, the earth's average temperature would be –240°C (–400°F), and life as we know it would not have arisen. Solar energy also helps recycle the chemicals we and other organisms need to stay alive and healthy (Figure 1-4).

Broadly defined, **solar energy** includes both direct energy from the sun and several forms of energy produced indirectly by the sun's energy: wind, falling and flowing water (hydropower), and biomass (solar energy converted to chemical energy stored in the chemical bonds of organic compounds in trees and other plants that can be burned to provide heat).

The remaining 1% of the energy we use to supplement the solar input is either *commercial energy* sold in the marketplace or *noncommercial energy* used by people who gather fuelwood, dung, and crop wastes for their own use. In MDCs about 90% of the commercial energy used is supplied by nonrenewable sources of energy (Figure 5-1); about 85% is provided by burning

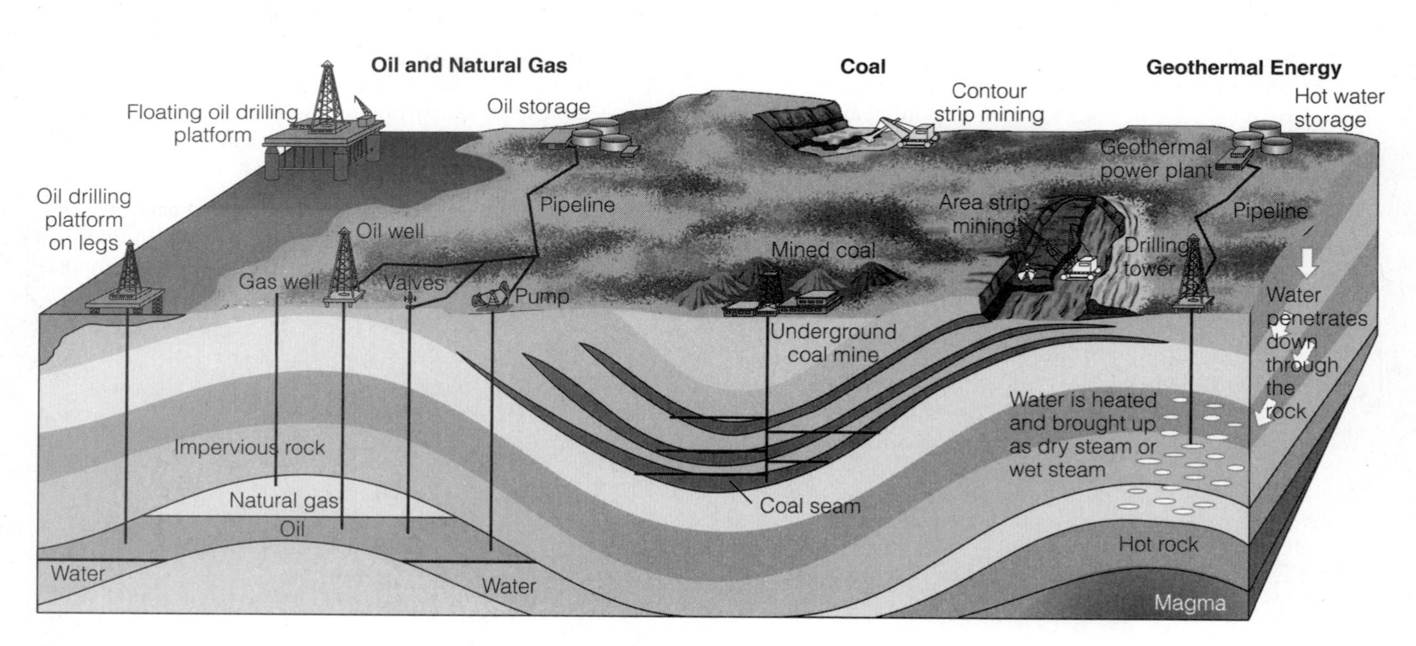

Figure 5-1 Important energy resources from the earth's crust include oil and natural gas, coal, and geothermal energy. Uranium ore (not pictured) is also extracted from the crust and then processed to increase the concentration of one type of its atoms, called uranium-235, which can be used as a fuel in nuclear reactors to produce electricity.

Q: What percentage of the earth's species live in tropical forests?

three nonrenewable fossil fuels—oil (and gasoline made from it), coal, and natural gas—and 5% is from nuclear power. The remaining 10% is provided from renewable sources of energy, with 7% coming from hydropower, geothermal energy, and solar energy and 3% from burning biomass.

By contrast, LDCs get 41% of their energy from renewable sources—mostly from biomass burned for heating and cooking (35%) and from hydropower and geothermal and solar energy (6%). Within a few decades, one-fourth of the world's population in MDCs may face an oil shortage, but half the world's population in LDCs already faces a fuelwood (biomass) shortage.

The United States is the world's largest user (and waster) of energy. With only 4.7% of the population, it uses 25% of the commercial energy, mostly by getting 91% of its energy from nonrenewable sources of energy (84% from fossil fuels and 7% from nuclear power). The remaining 9% comes from renewable energy sources (4% from biomass and 5% from hydropower, geothermal energy, and solar energy).

ENERGY QUALITY **Energy quality** is a measure of a source of energy's ability to do useful work (Figure 5-2). **High-quality energy** is organized or concentrated and can perform a great deal of useful work. Examples of these useful sources of energy are electricity, coal, gasoline, concentrated sunlight, nuclei of certain types of uranium atoms used as fuel in nuclear power plants and nuclear bombs, and heat concentrated in fairly small amounts of matter so that its temperature is high.

By contrast, **low-quality energy** is disorganized or dispersed and has little ability to do useful work. An example is heat dispersed in the moving molecules of a large amount of matter, such as the atmosphere or a large body of water, so that its temperature is relatively low.

We use energy to accomplish certain tasks, each requiring a certain minimum energy quality (Figure 5-2). It makes sense to match the quality of an energy source to the quality of energy needed to perform a particular task, for this saves energy and usually money.

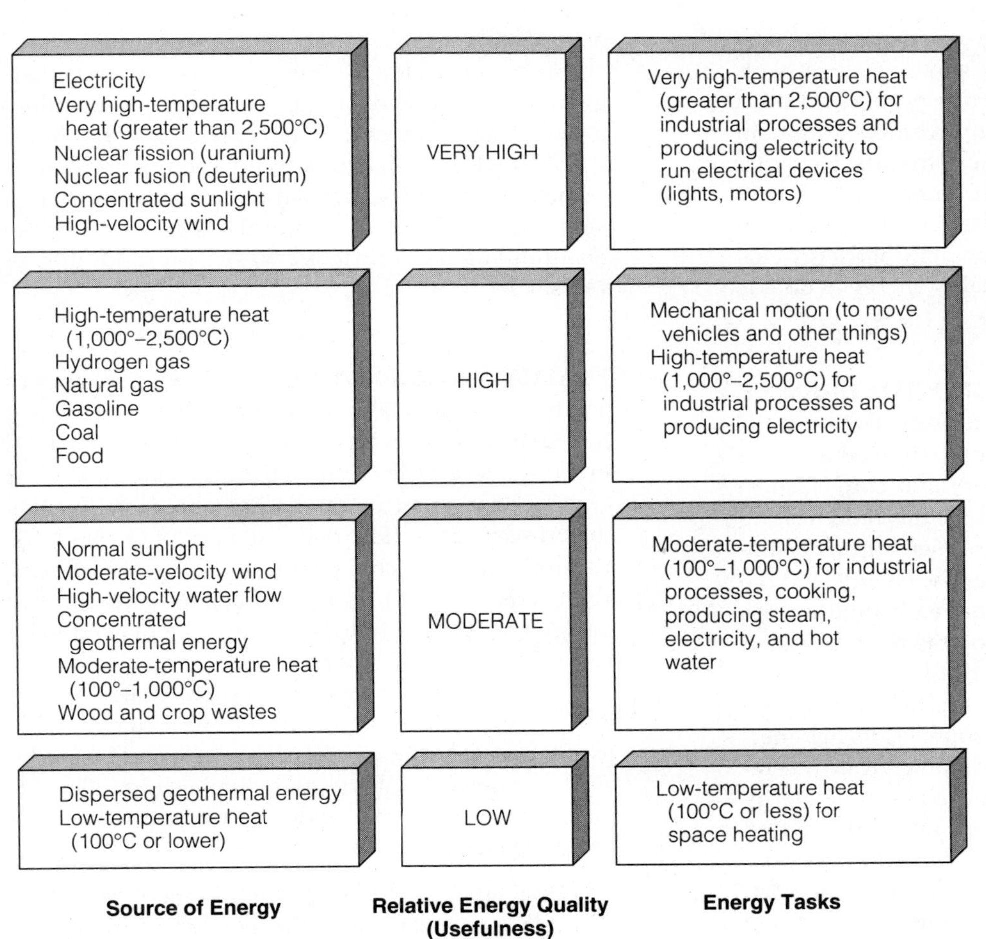

Figure 5-2 Generalized categories of the quality (or usefulness for performing various energy tasks) of different sources of energy. *High-quality energy* is concentrated and has great ability to perform useful work. *Low-quality energy* is dispersed and has little ability to do useful work. To avoid unnecessary energy waste, it's best to match the quality of an energy source with the quality of energy needed to perform a task.

A: At least 50% (some say 90%)

THE TWO IRONCLAD LAWS OF ENERGY Scientists have observed energy being changed from one form to another in millions of physical and chemical changes, but they have never been able to detect any energy being either created or destroyed. This summary of what happens in nature is called the **law of conservation of energy**, also known as the **first law of energy** or the **first law of thermodynamics**.

Because the first law of energy states that energy can be neither created nor destroyed, you might think that there will always be enough energy; yet, if you fill a car's tank with gasoline and drive around, or if you use a flashlight battery until it is dead, you have lost something. If it isn't energy, what is it? The answer is energy quality.

Countless experiments have shown that when energy is changed from one form to another, there is always a decrease in energy quality. Some of the useful energy is always degraded to lower-quality, more-dispersed, less-useful energy (usually heat, which flows into the environment and is dispersed by the random motion of air or water molecules). This summary of what we always find happening in nature is called the **second law of energy**, or the **second law of thermodynamics**. No one has ever found a violation of these two fundamental scientific laws.

The second law of energy means that *we can never recycle or reuse high-quality energy to perform useful work.* Once the concentrated energy in a serving of food, a liter of gasoline, a lump of coal, or a chunk of uranium is released, it is degraded to low-quality heat that becomes dispersed in the environment. We can heat air or water at a low temperature and upgrade it to high-quality energy, but the second law of energy tells us that it will take more high-quality energy to do this than we get in return.

NET ENERGY: THE ONLY ENERGY THAT REALLY COUNTS It takes energy to get energy. For example, crude oil must be found, pumped up from beneath the ground, transported to a refinery and converted to useful fuels (such as gasoline, diesel fuel, and heating oil), then transported to users, and then burned before it is useful to us. All of these steps use energy, and the second law of energy tells us that each time we use energy to perform a task, some of it is always wasted and degraded to low-quality energy.

The amount of high-quality or usable energy from a given quantity of an energy resource is its **net energy**—the total useful energy available from the resource over its lifetime, minus the amount of energy used (the first law of energy), automatically wasted (the second law of energy), and unnecessarily wasted in finding, processing, concentrating, and transporting it to users. For example, if 8 units of fossil fuel energy are needed to find, extract, process, upgrade, transport, and supply 10 units of nuclear, solar, or additional fossil fuel energy to users, the net energy gain is only 2 units of energy. Thus, net energy—not the total energy available from a particular energy resource—is the only energy that really counts.

We can look at this concept in a different way—as the ratio of useful energy produced to the useful energy used to produce it. In the example just given, the *net energy ratio* would be 10/8, or approximately 1.2. The higher the ratio, the greater the net energy yield. When the ratio is less than 1, there is a net energy loss over the lifetime of the system.

Figure 5-3 provides estimated net energy ratios for various systems of space heating, high-temperature heat for industrial processes, and transportation. Currently, oil has a relatively high net energy ratio because much of it comes from large, accessible deposits such as those in Saudi Arabia and other parts of the Middle East. When those sources are depleted, however, the net energy ratio of oil will decline and prices will rise. Then more money and more high-quality fossil fuel will be needed to find, process, and deliver new oil—from widely dispersed small deposits and deposits buried deep in the earth's crust or located in remote areas like Alaska, the Arctic, and the North Sea.

Conventional nuclear energy has a low net energy ratio because large amounts of energy are required to extract and process uranium ore, to convert it into a usable nuclear fuel, and to build and operate power plants. Energy is also needed to dismantle the plants after their 25–30 years of useful life and to store the resulting highly radioactive wastes for thousands of years.

EVALUATING ENERGY RESOURCES The types of energy we use and how we use them are the major factors determining both our quality of life and how much we abuse the earth's life-support system. Our current dependence on nonrenewable fossil fuels is the primary cause of air and water pollution, land disruption, and projected global warming and climate change (Section 6-2); and affordable oil will probably be depleted within 40 to 80 years and will need to be replaced by other alternatives.

Past experience shows that it usually takes at least 50 years and huge investments to phase in new energy alternatives. Thus, we must plan for and begin the shift to a new mix of energy resources now. This involves answering the following questions for each energy alternative:

- How much of this energy source might be available in the near future (the next 15 years), the

Q: What percentage of tropical forest plants have been studied for their possible use as human resources?

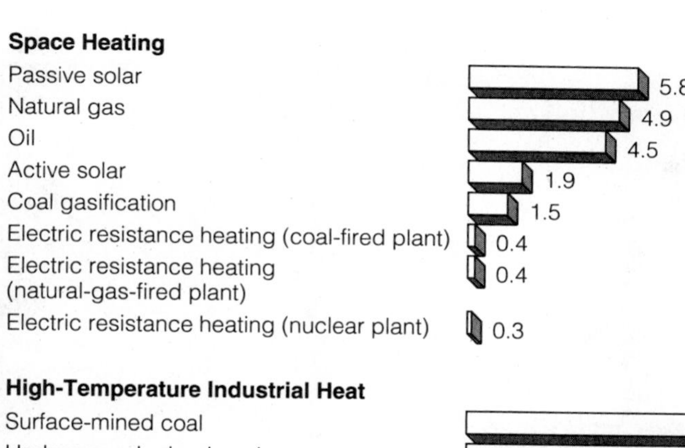

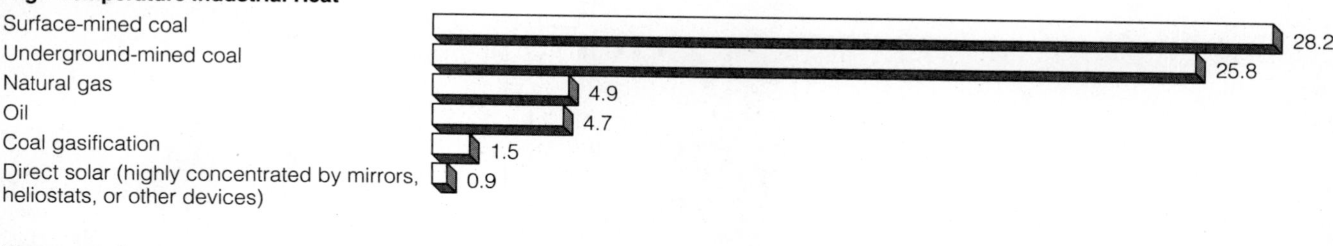

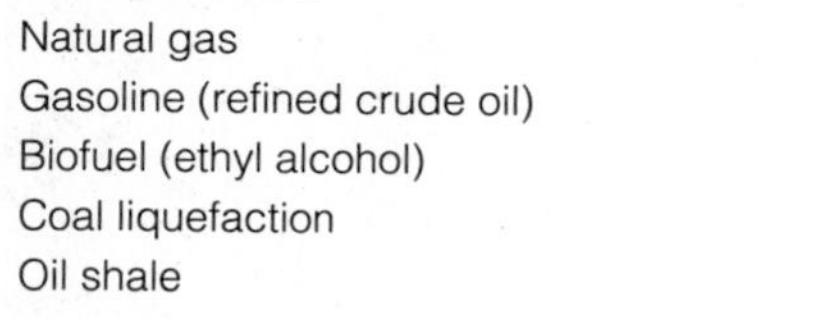

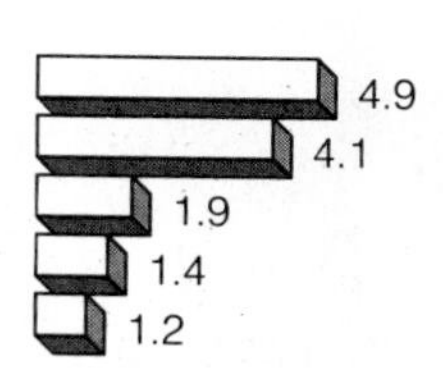

Figure 5-3 Net energy ratios for various energy systems over their estimated lifetimes. (Data from Colorado Energy Research Institute, *Net Energy Analysis*, 1976; and Howard T. Odum and Elisabeth C. Odum, *Energy Basis for Man and Nature*, 3rd ed., New York: McGraw-Hill, 1981)

intermediate future (the next 30 years), and for the long term (the next 50 years)?

- What is its net energy yield?
- How much will it cost to develop, phase in, and use this energy source?
- How will extracting, transporting, and using this energy source affect the environment (Figure 1-9)?
- What will this energy choice do to help sustain the earth for us, for future generations, and for the other species living on this planet?

5-2 Improving Energy Efficiency

DOING MORE WITH LESS You may be surprised to learn that *84% of all commercial energy used in the United States is wasted* (Figure 5-4). About 41% of this energy is wasted automatically because of the degradation of energy quality imposed by the second law of energy. However, about 43% is wasted unnecessarily, mostly by using fuel-wasting motor vehicles, furnaces, and other devices—and by living and working in leaky, poorly insulated buildings. People in the United States unnecessarily waste as much energy as two-thirds of the world's population consumes.

The easiest, fastest, and cheapest way to get more energy with the least environmental impact is to eliminate much of this energy waste. One way to do this is to reduce energy consumption. Examples include walking or biking for short trips, using mass transit, putting on a sweater instead of turning up the thermostat, and turning off unneeded lights.

Another way is to increase the efficiency of the energy conversion devices we use. **Energy efficiency** is the percentage of total energy input that does useful work (is not converted to low-quality, essentially useless heat) in an energy conversion system. We can save energy and money by buying the most energy-efficient home heating systems, water heaters, cars, air conditioners, refrigerators, and other household appliances available and by supporting research to find even more energy-efficient devices. The energy-efficient models may cost more, but in the long run they usually save money by having a lower *life-cycle cost*: initial cost plus lifetime operating costs.

A: 1%

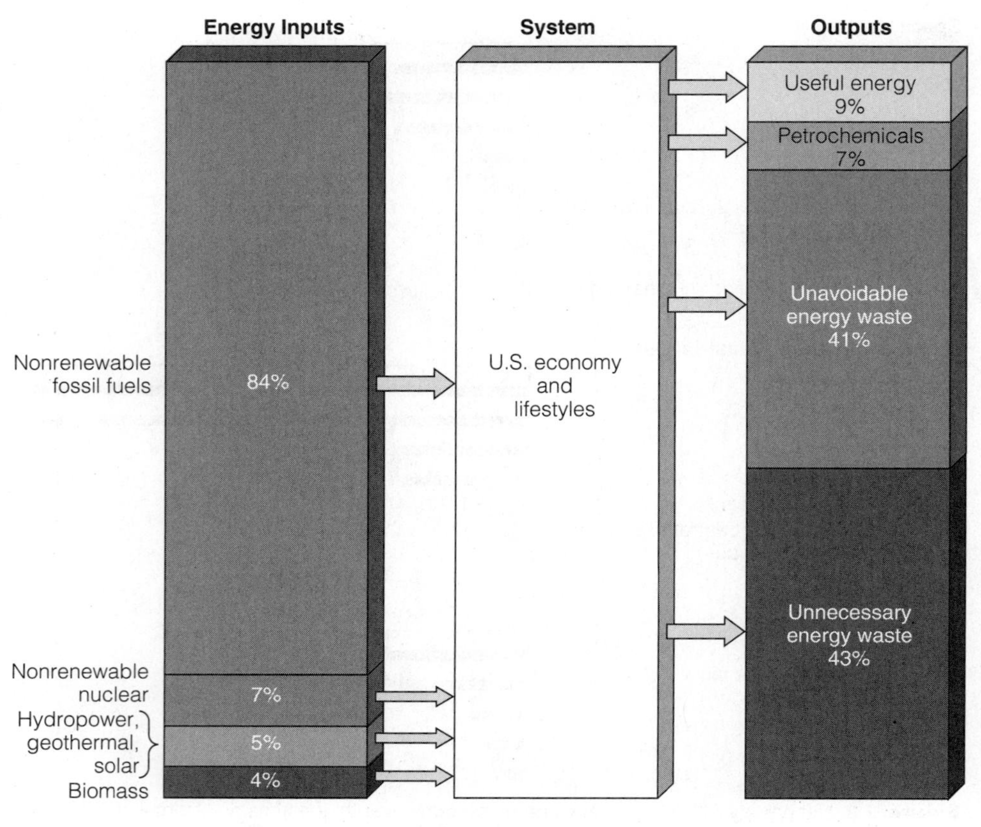

Figure 5-4 Flow of commercial energy through the U.S. economy. Note that only 16% of all commercial energy used in the United States ends up performing useful tasks or is converted to petrochemicals. The rest either is automatically and unavoidably wasted because of the second law of energy (41%) or is wasted unnecessarily (43%).

The net efficiency of the entire energy delivery process for a space heater, water heater, or car is determined by the efficiency of each step in the energy-conversion process. For example, the sequence of energy-using (and energy-wasting) steps involved in using electricity produced from fossil or nuclear fuels is

extraction → transportation → processing → transportation to power plant → electric generation → transmission → end use.

Figure 5-5 shows the net energy efficiency for heating a well-insulated home **(1)** with electricity produced at a nuclear power plant, transported by wire to the home, and converted to heat (electric resistance heating), and **(2)** passively with an input of direct solar energy through windows facing the sun, with heat stored in rocks or water for slow release. This analysis shows that the process of converting the high-quality energy in nuclear fuel to high-quality heat at several thousand degrees, converting this heat to high-quality electricity, and then using the electricity to provide low-quality heat for warming a house to only about 20°C (68°F) is extremely wasteful of high-quality energy. Burning coal (or any fossil fuel) at a power plant to supply electricity for space heating is also inefficient. By contrast, it is much less wasteful to collect solar energy from the environment, store the resulting heat in stone or water, and—if necessary—raise its temperature slightly to provide space heating or household hot water.

Physicist and energy expert Amory Lovins points out that using high-quality electrical energy to provide low-quality heating for living space or household water is like using a chain saw to cut butter or a sledgehammer to kill a fly. As a general rule, he suggests that we not use high-quality energy to do a job that can be done with lower-quality energy. The logic of his point is illustrated by looking at the prices for providing heat using various fuels. In 1991, the average price of obtaining 250,000 kilocalories (1 million Btu) for heating either space or water in the United States was $6.05 using natural gas, $7.56 using kerosene, $9.30 using oil, $9.74 using propane, and $24.15 using electricity. As these numbers suggest, if you don't mind throwing away hard-earned dollars, then use electricity to heat your house and bath water.

REDUCING ENERGY WASTE: AN ECONOMIC AND ENVIRONMENTAL OFFER WE DARE NOT REFUSE

Reducing energy waste is one of the planet's best and most important bargains. Consider the following benefits of reducing energy waste:

- *Making nonrenewable fossil fuels last longer.*
- *Buying time to phase in renewable energy resources.*

Q: How much of the world's area of tropical forests is managed sustainably?

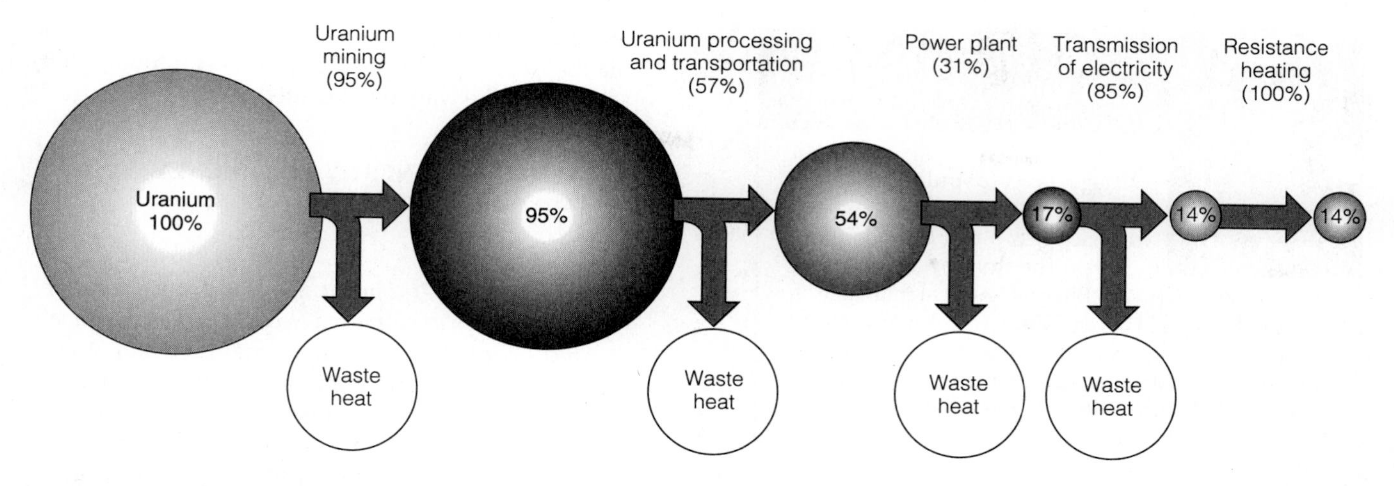

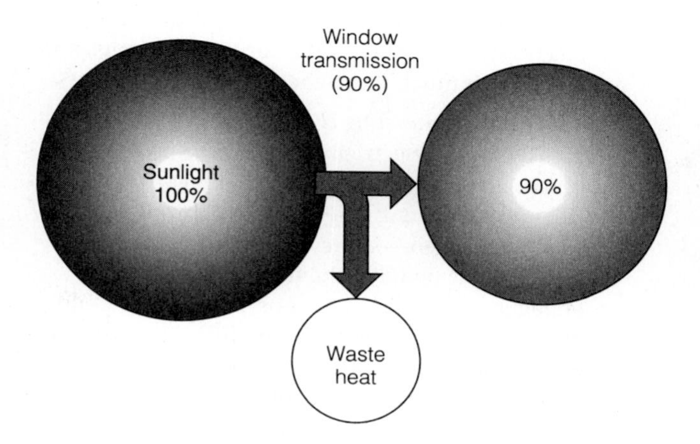

Figure 5-5 Comparison of net energy efficiency for two types of space heating. The cumulative net efficiency is obtained by multiplying the percentage shown inside the circle for each step by the energy efficiency for that step (shown in parentheses). Usually, the greater the number of steps in an energy conversion process, the lower its net energy efficiency. About 86% of the energy used to provide space heating by electricity produced at a nuclear power plant is wasted. By contrast, with passive solar heating only about 10% of incoming solar energy is wasted.

- *Decreasing dependence on oil imports* (almost 51% in the United States in 1993).
- *Lessening the need for military intervention in the oil-rich but potentially unstable Middle East.*
- *Reducing environmental damage* because less of each energy resource would provide the same amount of useful energy.
- *Providing the cheapest and quickest way to slow projected global warming by reducing emissions of heat-trapping carbon dioxide when fossil fuels are burned* (Section 6-2).
- *Saving money, providing more jobs, and promoting more economic growth per unit of energy than other alternatives.* According to energy expert Amory Lovins, if the world really got serious about improving energy efficiency, we could save $1 trillion per year—about 5% of the gross global product. A 1993 study by economists estimated that a full-fledged energy-efficiency program could produce 1.3 million jobs in the United States by 2010.
- *Improving competitiveness in the international marketplace.* Currently, the United States spends about 11% of its GNP on energy, whereas Japan spends only 5%. That gives Japanese goods a cost advantage.

Energy analysts and the U.S. Office of Technology Assessment estimate that fully implementing *existing* energy efficiency technologies could save the United States four times as much electricity as all U.S. nuclear power plants now produce, at about one-seventh the cost of just running them, even if it cost nothing to build them. Further, such a program would save more than 40 times as much oil as *might* be under Alaska's Arctic National Wildlife Refuge (Case Study, p. 76), at roughly one-tenth the cost of drilling for it.

Japan and many western-European countries are leading the world in the *energy-efficiency revolution*. A

A: 0.1%

Saving Energy, Money, and Jobs in Osage, Iowa

SOLUTIONS

Osage, Iowa (population about 4,000), has become the energy efficiency capital of the United States. The story began in 1974, when easy-going Wes Birdsall, general manager of Osage Municipal Gas and Electric Company, started going door-to-door preaching energy conservation to help his community deal with the energy crisis of 1973. The utility would also save money by not having to add new electrical generating facilities.

Wes started his crusade by telling homeowners about the importance of insulating walls and ceilings and of plugging leaky windows and doors. These repairs provided jobs for people selling and installing insulation, caulking, and energy-efficient windows.

He also advised people to replace their incandescent light bulbs with more efficient fluorescent bulbs (Figure 5-6, p. 86) and to turn down the temperature on water heaters and wrap them with insulation, an economic boon to the local hardware and lighting stores. Wes also suggested saving water and fuel by installing low-flow shower heads.

Wes then stepped up his campaign, offering to give every building in town a free thermogram—an infrared scan that shows where heat escapes. When people could see the energy (and money) hemorrhaging out of their buildings, they took action to plug these leaks, again helping the local economy and saving themselves money.

Next Wes announced that no new houses could be hooked up to the company's natural gas line unless they met minimum energy-efficiency standards.

Since 1974, the town has cut its natural gas consumption by 45%, no mean feat in a place where winter temperatures can plummet to –103°C (–80°F). In addition, the utility company saved enough money to prepay all its debt, accumulate a cash surplus, and cut inflation-adjusted electricity rates by a third (which attracted two new factories). Furthermore, each household saves more than $1,000 per year. This money supports jobs, and most of it circulates in the local economy. Before this energy-efficiency revolution, about $1.2 million a year went out of town—usually out of state—to buy energy.

Osage's success in making energy efficiency a way of life earned the town a National Environmental Achievement Award in 1991. What are your local utility and community doing to improve energy efficiency and stimulate the local economy?

few places in the United States, such as Davis, California (Solutions, p. 28) and Osage, Iowa (Solutions, left), are leading the way in the United States.

WAYS TO SAVE ENERGY (AND MONEY)

Industry

- *Cogeneration—the production of two useful forms of energy (such as steam and electricity) from the same fuel source.* Waste heat from coal-fired and other industrial boilers can be used to produce steam that spins turbines and generates electricity at half the cost of buying it from a utility company. By using the electricity or selling it to the local power company for general use, a plant can save energy and money. Cogeneration, widely used in western Europe for years, could produce more electricity than all U.S. nuclear power plants, and could do it much more cheaply.
- *Replace energy-wasting electric motors.* About 60–70% of the electricity used in U.S. industry drives electric motors. Most of them are run at full speed, but with their output "throttled" to match their task—somewhat like driving a car with the gas pedal to the floor and the brake engaged. According to Amory Lovins, it would be cost-effective to scrap virtually all such motors and replace them with adjustable-speed drives; within a year the costs would be paid back.
- *Switch to high-efficiency lighting.*
- *Use computer-controlled energy management systems* to turn off lighting and equipment in nonproduction areas and to make adjustments during periods of low production.

Transportation

- *Increase the fuel efficiency of motor vehicles.* Between 1973 and 1985 the average fuel efficiency doubled for new American cars and rose 54% for all cars on the road, but it has risen only slightly since then. According to the U.S. Office of Technology Assessment, existing technology could be used to raise the fuel efficiency of the entire U.S. automotive fleet to 15 kilometers per liter (35 miles per gallon) by 2010, to eliminate oil imports, and to save more than $50 billion per year in fuel costs. Buyers of gas-miser cars would get back any extra purchase costs—probably about $500 per car—in fuel savings in about a year (Solutions, p. 85).
- *Shift more freight from trucks and planes to more energy-efficient trains and ships.*

Q: How many people cannot find or buy enough fuelwood to meet their basic needs?

Buildings

- *Build more superinsulated houses.* Such houses are so heavily insulated and sufficiently airtight that heat from direct sunlight, appliances, and human bodies warms them, with little or no need for a backup heating system. An air-to-air heat exchanger prevents buildup of indoor air pollution. Although such houses typically cost 5% more to build than conventional houses of the same size, this extra cost is paid back by energy savings within five years and can save a homeowner $50,000–$100,000 over a 40-year period.
- *Improve the energy efficiency of existing houses by adding insulation, plugging leaks, and installing energy-saving windows.* One-third of the heat in U.S. homes and buildings escapes through closed windows—an amount equal to the energy in all the oil flowing through the Alaskan pipeline every year. Superinsulating windows now available lose no more heat than a well-insulated exterior wall. They pay for themselves in lower fuel bills within two to five years, and then save money every year thereafter.
- *Use the most energy-efficient ways to heat houses.* The most energy-efficient ways to heat homes are to build a superinsulated house, to use passive solar heating, and to use high-efficiency (85–98% efficient) natural-gas furnaces. The most wasteful and expensive way is to use electric resistance heating with electricity produced by a nuclear (Figure 5-5) or a coal-fired power plant.
- *Use the most energy-efficient ways to heat household water.* An efficient method is to use tankless instant water heaters (about the size of a bookcase stereo speaker, widely used in many parts of Europe) fired by natural gas or liquefied petroleum gas (LPG). Such devices heat the water instantly as it flows through a small burner chamber, and they provide hot water only when (and as long as) it is needed. A well-insulated, conventional natural gas or LPG water heater is also fairly efficient (although all conventional natural gas and electric resistance heaters keep a large tank of water hot all day and night and can run out after a long shower or two). The least efficient and most expensive way to heat water for washing and bathing is to use electricity produced by any type of power plant.
- *Set higher energy-efficiency standards for buildings.* Building codes can be changed to require that all new houses use 80% less energy than conventional houses of the same size, as has been done in Davis, California (Solutions, p. 28). Because of tough energy-efficiency standards, the average Swedish home consumes about one-third as much energy as an average American home of the same size.

SOLUTIONS

Gas Sippers and Electric Cars

Many people believe that fuel-efficient cars will take decades to develop and will be sluggish, small, and unsafe. Wrong! Since 1985 at least 10 companies have had nimble and peppy prototype cars that meet or exceed current safety and pollution standards, with fuel efficiencies of 29–59 kilometers per liter (67–138 miles per gallon). If such cars were mass-produced, their slightly higher costs would be more than offset by their fuel savings.

We can have roomy, peppy, safe gas sippers, but only if consumers begin demanding them. In 1992 the more than 25 car models that got at least 17 kpl (40 mpg) made up only 5% of U.S. car sales (mostly because of low gasoline prices).

Most environmentalists believe that such cars will not be widely used—and much more efficient ones produced—without significant, government-mandated improvements in fuel efficiency and greatly increased gasoline taxes (coupled with tax relief for the poor and the lower-middle class). They argue that both of these actions will eventually save the country and consumers large amounts of money.

Electric cars might also help reduce dependence on oil, especially for urban commuting and short trips. All major U.S. car companies have prototype electric cars and minivans, some expected to be available by 1995. They are extremely quiet, need little maintenance, and produce no air pollution, except indirectly from the generation of electricity needed to recharge their batteries. If solar cells could be used for recharging, this environmental impact would be eliminated.

On the negative side, the batteries in current electric cars must be replaced about every 40,000 kilometers (25,000 miles) at a cost of at least $1,500. This requirement, and the electricity costs for daily recharging, mean double the operating costs of gasoline-powered cars. If longer-lasting batteries that hold a higher charge density and last at least 160,000 kilometers (100,000 miles) can be developed, operating costs would be reduced and performance would increase.

A: 1.5 billion—almost one person out of every three on Earth

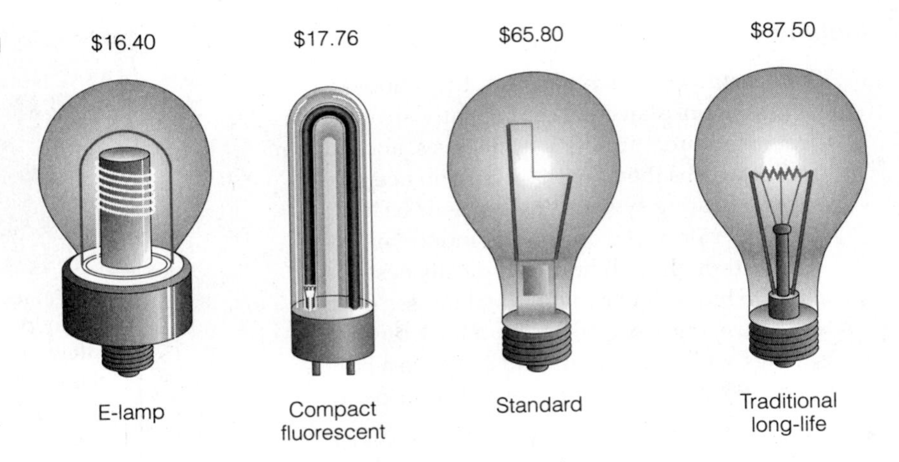

Figure 5-6 Light up your life, help the earth, and save money by using energy-efficient bulbs: cost of electricity for comparable light bulbs used for 10,000 hours. Because conventional incandescent bulbs are only 5% efficient and last only 750–1,500 hours, they waste enormous amounts of energy and money, and they add to the heat load of houses during hot weather. (Data from Electric Power Research Institute)

- *Buy the most energy-efficient appliances and lights** (Figure 5-6). If the most energy-efficient lights and appliances now available were installed in all U.S. homes over the next 20 years, the savings in energy would equal the estimated energy content of Alaska's entire North Slope oil fields. Replacing a standard incandescent bulb with an energy-efficient compact fluorescent or E-lamp bulb saves about $50—or $1,250 for a typical house with 25 light bulbs.
- *Give rebates or tax credits for building energy-efficient buildings, for improving the energy efficiency of existing buildings, and for buying high-efficiency appliances and equipment.*

5-3 Direct Use of Renewable Solar Energy

THE SOLAR AGE: BUILDING A SOLAR ECONOMY

About 92% of the known reserves and potentially available energy resources in the United States are *renewable* energy resources: sun, wind, flowing water, biomass, and the earth's internal heat. The other 8% of potentially available domestic energy resources are nonrenewable coal (5%), oil (2.5%), and uranium (0.5%). Developing the mostly untapped renewable energy resources could meet 50–80% of projected U.S. energy needs by 2030 or sooner and could meet virtually all energy needs if coupled with improvements in energy efficiency.

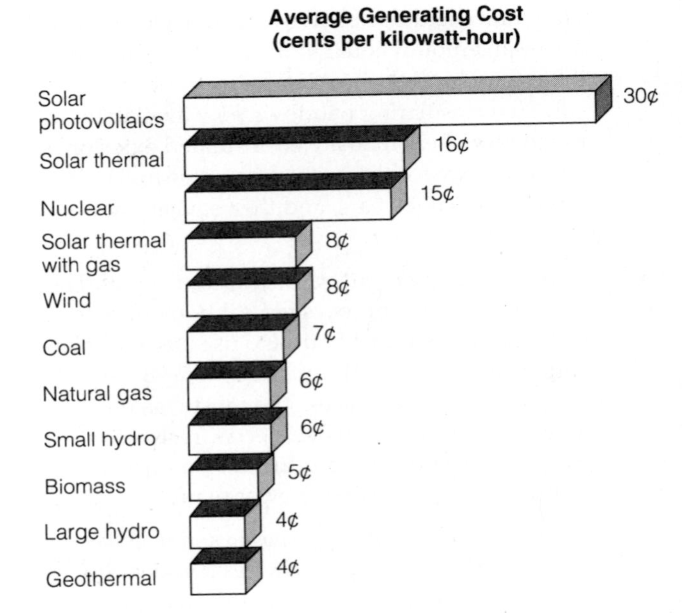

Figure 5-7 Generating costs in 1989 of electricity per kilowatt-hour by various technologies. By 2000, costs per kilowatt-hour for wind are expected to fall to 4–5¢, for solar thermal with gas assistance to 6¢, and for solar photovoltaic to 10¢. Costs for other technologies are expected to remain about the same. (Data from U.S. Department of Energy, Council for Renewable Energy Education, and Investor Responsibility Research Center)

Developing these resources would save money, create jobs, eliminate the need for oil imports, cause less pollution and environmental damage per unit of energy used, and increase economic, environmental, and military security (by replacing strategically vulnerable large coal and nuclear plants with a dispersed mix of smaller energy-producing facilities). In the United States, geothermal plants, wood-fired (biomass) plants, hydropower plants, wind farms, and solar thermal plants with natural gas backup can already produce

*Each year the American Council for an Energy-Efficient Economy (ACEEE) publishes a list of the most energy-efficient major appliances mass-produced for the U.S. market. To obtain a copy, send $3 to the council at 1001 Connecticut Ave. N.W., Suite 530, Washington, DC 20036. Each year they also publish *A Consumer Guide to Home Energy Savings*, available in bookstores or from the ACEEE for $8.95.

Q: By 2000, how many people may not be able to get enough fuelwood?

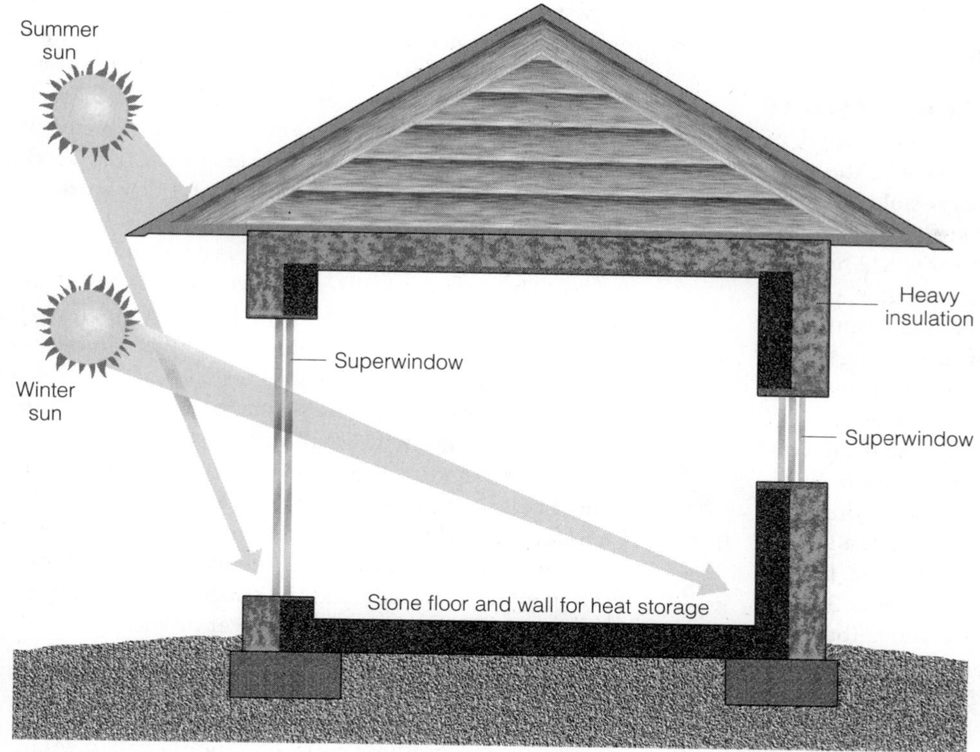

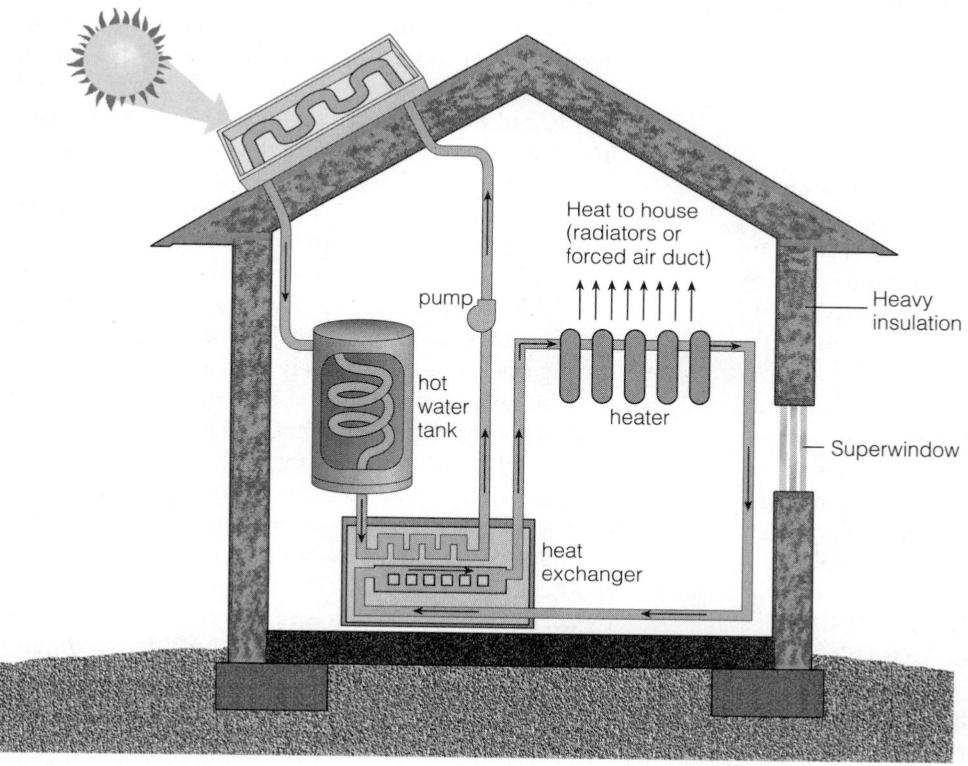

Figure 5-8 Passive and active solar heating for a home.

A: 2.7 billion

electricity more cheaply than can new nuclear power plants (Figure 5-7), and with far fewer federal subsidies.

HEATING HOUSES AND WATER WITH SOLAR ENERGY Buildings and water can be heated by solar energy using two methods: passive heating and active heating (Figure 5-8). A **passive solar heating system** captures sunlight within a structure and converts it into low-temperature heat for space heating. With available and developing technologies, passive solar designs can provide at least 80% of a building's heating needs and at least 60% of its cooling needs. Roof-mounted passive solar water heaters can supply all or most of the hot water for a typical house.

On a life-cycle cost basis, a good passive solar and superinsulated design is the cheapest way to heat a home or a small building in regions where sunlight is available more than 60% of the time. Such a system usually adds 5–10% to the construction cost, but the life-cycle cost of operating such a house is 30–40% lower. Most passive solar systems require that windows and shades be opened or closed to regulate heat flow and distribution, but this can be done by inexpensive microprocessors.

In an **active solar heating system**, collectors absorb solar energy, and a fan or a pump supplies part of a building's space-heating or water-heating needs (Figure 5-8). Active solar collectors can also supply hot water. With current technology, active solar systems usually cost too much for heating most homes and small buildings, but improved designs and mass-production techniques could change that.

Solar energy used for low-temperature heating of buildings, whether collected actively or passively, is free, and the net energy yield is moderate (active) to high (passive). Both active and passive technologies are well developed and can be installed quickly. No heat-trapping carbon dioxide (Section 6-2) is added to the atmosphere, and environmental impacts from air and water pollution are low. Land disturbance is also minimal because passive systems are built into structures and active solar collectors are usually placed on rooftops. Owners of passive and active solar systems also need "solar rights" laws to prevent others from building structures that block their access to sunlight.

USING SOLAR ENERGY TO GENERATE HIGH-TEMPERATURE HEAT AND ELECTRICITY
Several systems collect and concentrate enough solar energy to generate electricity and high-temperature heat (Figure 5-9). The most promising approach to intensifying solar energy is *nonimaging optics*, under development since 1965 by American and Israeli scientists.

Because of their high efficiency and ability to generate extremely high temperatures, nonimaging optical concentrators may make solar energy practical for widespread industrial and commercial use within a decade. Giant arrays of these concentrators at solar power plants may power turbines to generate electricity, to produce hydrogen gas for fuel (Section 5-5), and to convert some hazardous wastes into less harmful substances. Inexpensive solar cookers can also be used to concentrate solar energy and cook food (Figure 5-9e), especially in rural villages in sunny LDCs.

The impact of solar power plants on air and water is low. They can be built in 1–2 years, compared to 5–15 years for coal-fired and nuclear power plants. Thus builders would save millions of dollars in interest on construction loans. Solar thermal power plants produce electricity almost as cheaply as new nuclear power plants; and, with small turbines burning natural gas as a backup, they can produce electricity at almost half the cost of nuclear power plants (Figure 5-7). Solar thermal power plants still use one-third less land area than a coal-burning plant (when the land used to extract coal is included), and 95% less land per kilowatt-hour than most hydropower projects.

PRODUCING ELECTRICITY FROM SOLAR CELLS
Solar energy can be converted directly into electrical energy by **photovoltaic cells**, commonly called **solar cells** (Figure 5-10)—an entirely different technology than active solar collectors, which yield heat, not electricity. Solar cells are reliable and quiet, have no moving parts, and should last 30 years or more if encased in glass or plastic. They can be installed quickly and easily, and maintenance consists of occasional washing to keep dirt from blocking the sun's rays. Small or large solar-cell packages can be built, and they can be expanded easily or moved as needed. Solar cells can be located in deserts and on marginal lands, alongside interstate highways, in yards, and on rooftops.

Solar cells produce no heat-trapping carbon dioxide (Section 6-2) during use. Air and water pollution during operation is extremely low, air pollution from their manufacture is low, and land disturbance is very low for roof-mounted systems. The net energy yield is fairly high and is increasing with new designs. By 2030, electricity produced by photovoltaic cells could drop to 4¢ per kilowatt-hour, making it fully cost-competitive. With an aggressive program starting now, solar cells could supply 17% of the world's electricity by 2010—as much as nuclear power does today, at a lower cost and at much lower risk; by 2050 that figure could reach 30% (50% in the United States).

There are some drawbacks, however. The current costs of solar-cell systems are high (Figure 5-7), but they should become competitive in 5–15 years and are already cost-competitive in some situations. Moderate levels of water pollution from chemical wastes introduced through the manufacturing process can be a problem in the absence of effective pollution controls.

Q: What percentage of the original old-growth forests in the United States have been cut?

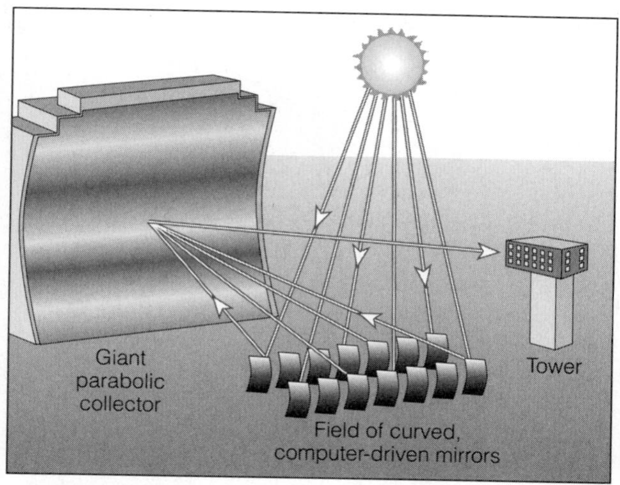

a. Solar Furnace

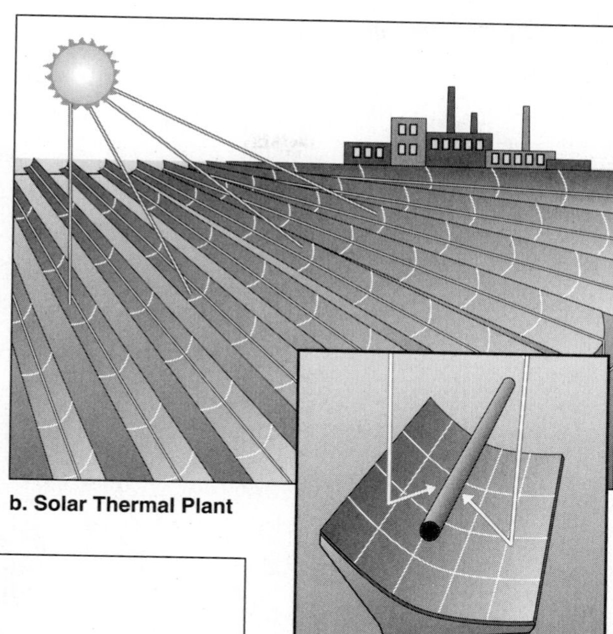

b. Solar Thermal Plant

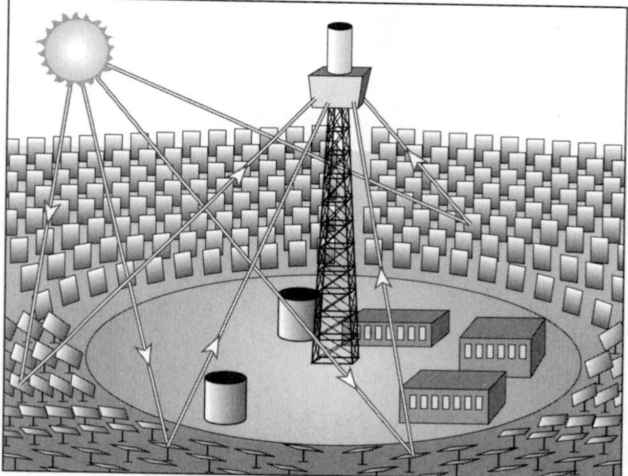

c. Solar Power Tower

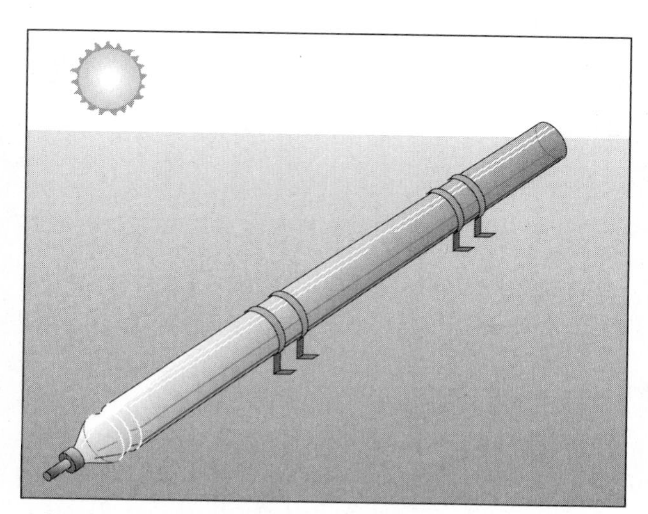

d. Nonimaging Optical Solar Concentrator

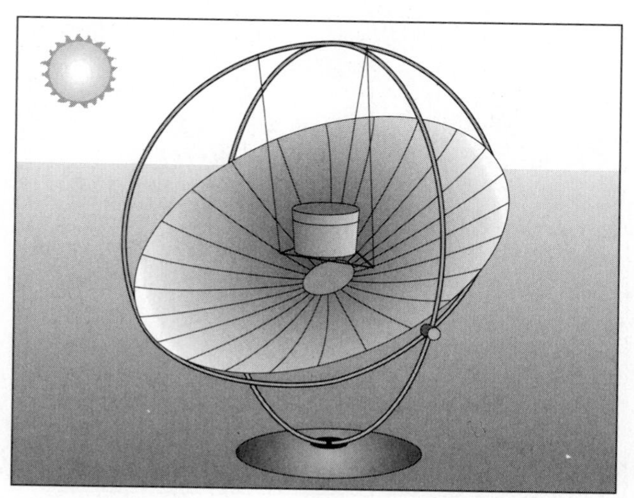

e. Solar Cooker

Figure 5-9 Several ways to collect and concentrate solar energy to produce high-temperature heat and electricity are in use. Today such plants are used mainly to supply reserve power for daytime peak electricity loads, especially in sunny areas with a large demand for air conditioning. Backed up by small natural-gas turbines, solar thermal plants occupying less than 1% of the Mojave Desert could probably supply Los Angeles with electricity. On an even economic playing field such plants could produce electricity more cheaply than a coal-burning or nuclear power plant.

A: 90–95%

Single Solar Cell

boron-enriched silicon
sunlight
junction
cell
phosphorus-enriched silicon
DC electricity

Array of Solar Cell Panels on a Roof

photovoltaic panels
power lines
panel wire
to breaker panel (inside house)
inverter (converts DC to AC)
battery bank (located in shed outside house, due to explosive nature of battery gases)

Panel of Solar Cells

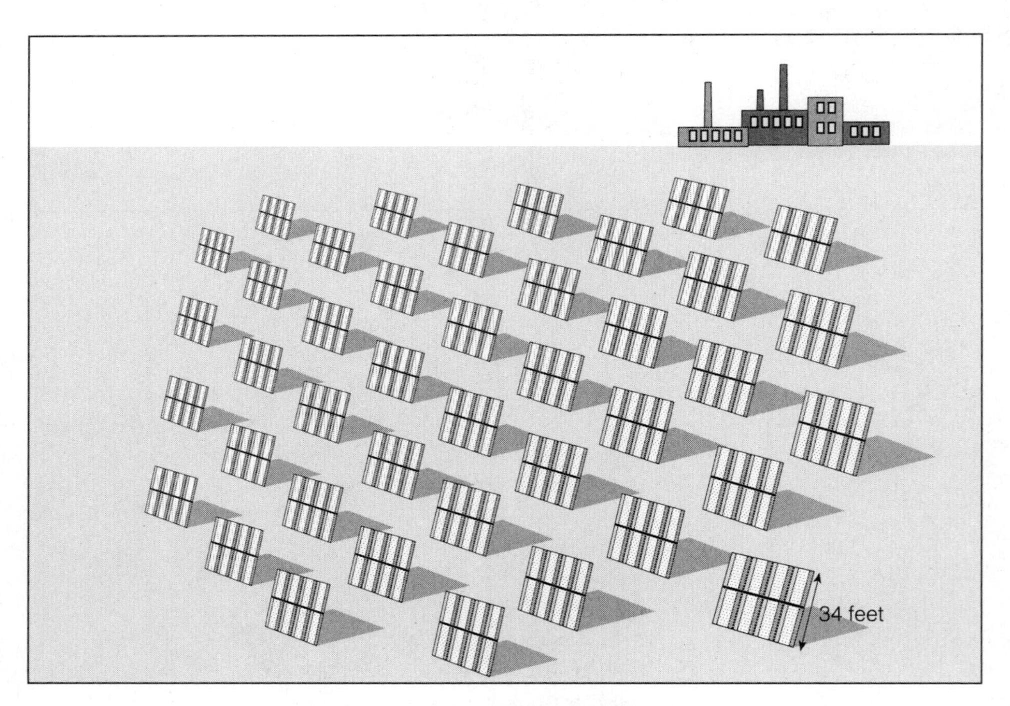

Figure 5-10 Use of photovoltaic (solar) cells to provide electricity for a house. Small and easily expandable arrays of such cells can provide electricity for villages throughout the world without the need for building large power plants and power lines. Massive banks of such cells can also produce electricity at small power plants. Today at least two dozen U.S. utility companies are using photovoltaic cells in their operations. As the price of such electricity drops, usage will increase dramatically.

Q: How much of the U.S. Forest Service budget is devoted to timber sales?

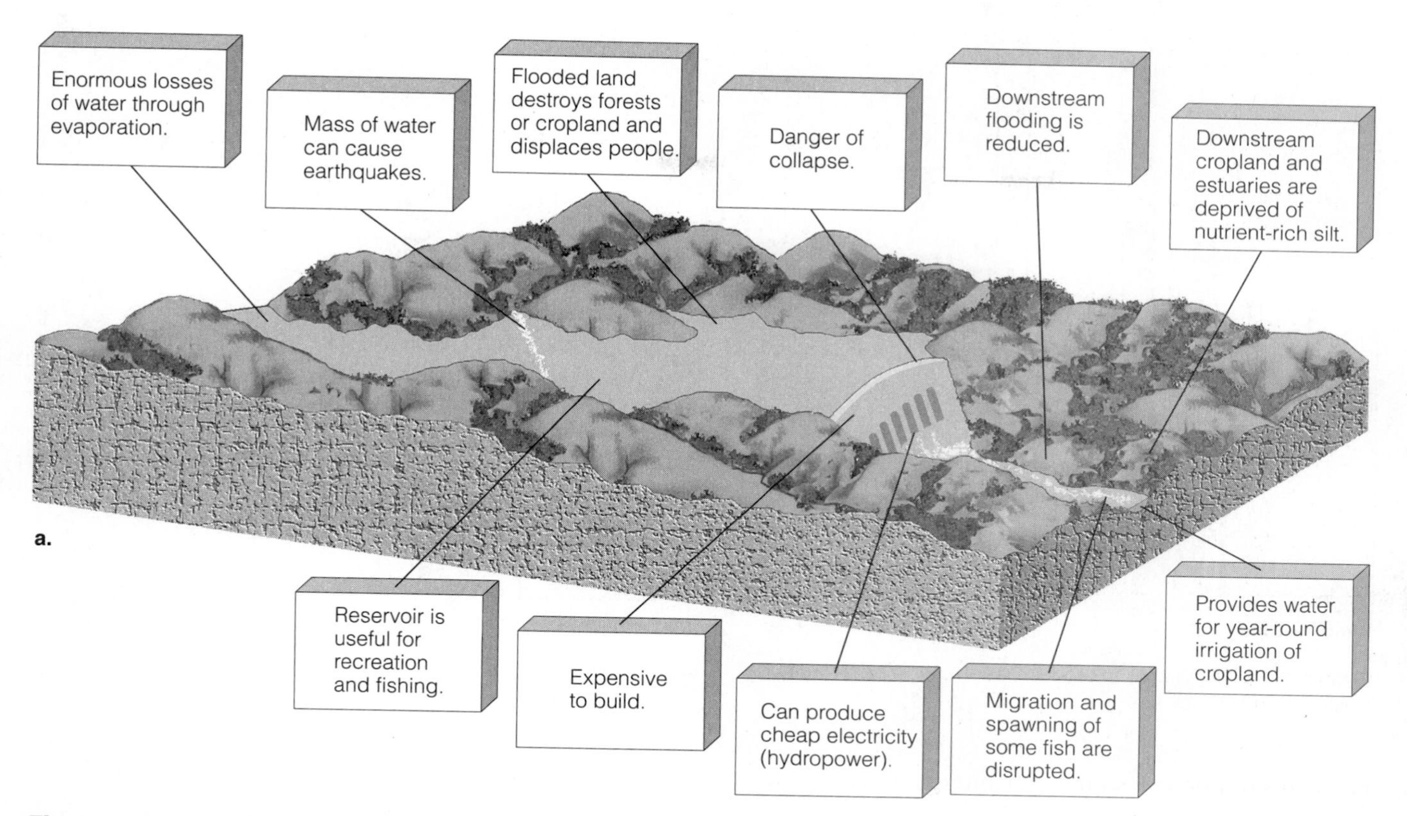

Figure 5-11 A depiction of a large hydropower project, showing advantages and disadvantages of large dams and reservoirs.

Also, some people find racks of solar cells on rooftops or in yards to be unsightly, but new thin and flexible rolls of solar cells already developed will eliminate this problem. Unless federal and private research efforts on photovoltaics are increased sharply, the United States will lose out on a huge global market (at least $5 billion per year by 2010) and may have to import photovoltaic cells from Japan, Germany, Italy, and other countries that have been investing heavily in this promising technology since 1980.

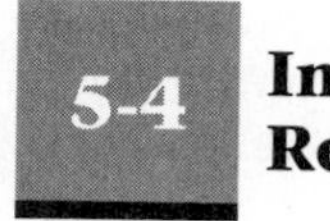

Indirect Use of Renewable Solar Energy

PRODUCING ELECTRICITY FROM FLOWING WATER In *large-scale hydropower projects*, high dams are built across large rivers to create large reservoirs (Figure 5-11). The stored water then flows through huge pipes at controlled rates, spinning turbines and producing electricity. In *small-scale hydropower projects*, a low dam with no reservoir (or only a small one) is built across a small stream. Because natural water flow generates the electricity, output of small systems can vary with seasonal changes in stream flow.

Hydroelectric power, or hydropower, supplies about 20% of the world's electricity and 5% of its total commercial energy. Hydropower supplies Norway with essentially all of its electricity, Switzerland with 74%, Canada with 70%, Austria with 67%, and LDCs with 50%. Large-scale development of hydropower is planned in many LDCs, including China, India, and Brazil. The United States is currently the world's largest producer of hydroelectricity, which supplies 10% of its electricity and 3–5% of all commercial energy. However, the era of large dams is ending in the United States because construction costs are high, few suitable sites are left, and environmentalists oppose many of the proposed projects because of their harmful effects (Figure 5-11).

Hydropower has a moderate to high net energy yield and fairly low operating and maintenance costs. Hydroelectric plants rarely need to be shut down, and they emit no heat-trapping carbon dioxide or other air pollutants during operation. They have life spans 2–10 times those of coal and nuclear plants. Large dams also help control flooding and supply a regulated flow of irrigation water to areas below the dam.

Electricity can also be produced from tides, ocean waves, and the heat stored in water of tropical oceans and in solar ponds; however, none of these are expect-

A: About 74%

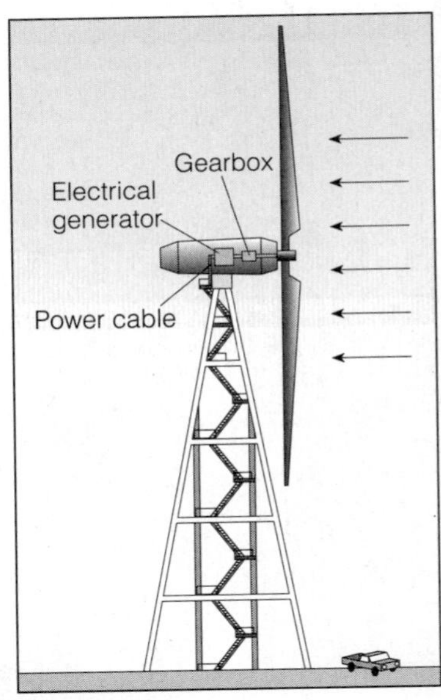

Wind Turbine

Wind Farm

Figure 5-12 Using wind to produce electricity. Wind turbines can be used individually or in clusters (wind farms).

ed to be important sources of energy in the near future.

PRODUCING ELECTRICITY FROM WIND In 1993 there were over 20,000 wind turbines worldwide, most grouped in clusters called *wind farms* (Figure 5-12). Most are in California (17,000 machines, which produce enough electricity to meet the residential needs of a city as large as San Francisco) and Denmark (which gets 2–3% of its electricity from wind turbines). Most are located in windy mountain passes and on ridges and along coastlines that generally have strong and steady winds. If wind farms were built on favorable sites in North Dakota, that state alone could supply 20–36% of the electricity currently used in the continental United States.

Wind power is a virtually unlimited source of energy at favorable sites, and large wind farms can be built in 6–12 months and then easily expanded as needed. With a moderate to fairly high net energy yield, these systems emit no heat-trapping carbon dioxide or other air pollutants during operation; they need no water for cooling, making them produce little air or water pollution; and they operate 95–98% of the time. The land under wind turbines can be used for grazing cattle and for other purposes, and the leases to use the land for wind turbines can provide extra income for farmers and ranchers. Wind power (with much lower subsidies) also has a significant cost advantage over nuclear power (Figure 5-7) and should become competitive with coal in many areas before 2000.

However, wind power is economical only in areas with steady winds. When the wind dies down, backup electricity from a utility company or from an energy storage system becomes necessary. Backup power could also be provided by linking wind farms with a solar-cell or hydropower system or with efficient natural-gas turbines. Other drawbacks to wind farms include visual pollution and noise, although these disadvantages can be overcome with improved design and placement in isolated areas. Large wind farms might also interfere in certain areas with the flight patterns of migratory birds and can kill large birds of prey (especially hawks, falcons, and eagles) that prefer to hunt along the same ridge lines that are ideal for wind turbines.

Wind power experts project that by the middle of the twenty-first century wind power could supply more than 10% of the world's electricity and 10–25% of the electricity used in the United States. Danish companies, with the aid of tax incentives and low-interest loans from their government, have taken over the lion's share of the global market for manufacturing wind turbines. European governments are currently spending 10 times more for wind energy research and development than the U.S. government and plan to produce almost twice as much electricity from the wind by 2000 as the United States.

BIOMASS: A VERSATILE FUEL *Biomass* is organic matter produced directly (in producers) by solar energy through photosynthesis and indirectly by consumers feeding on producers. It includes wood, agri-

Q: Between 1978 and 1992, how much money did the Forest Service lose on timber sales?

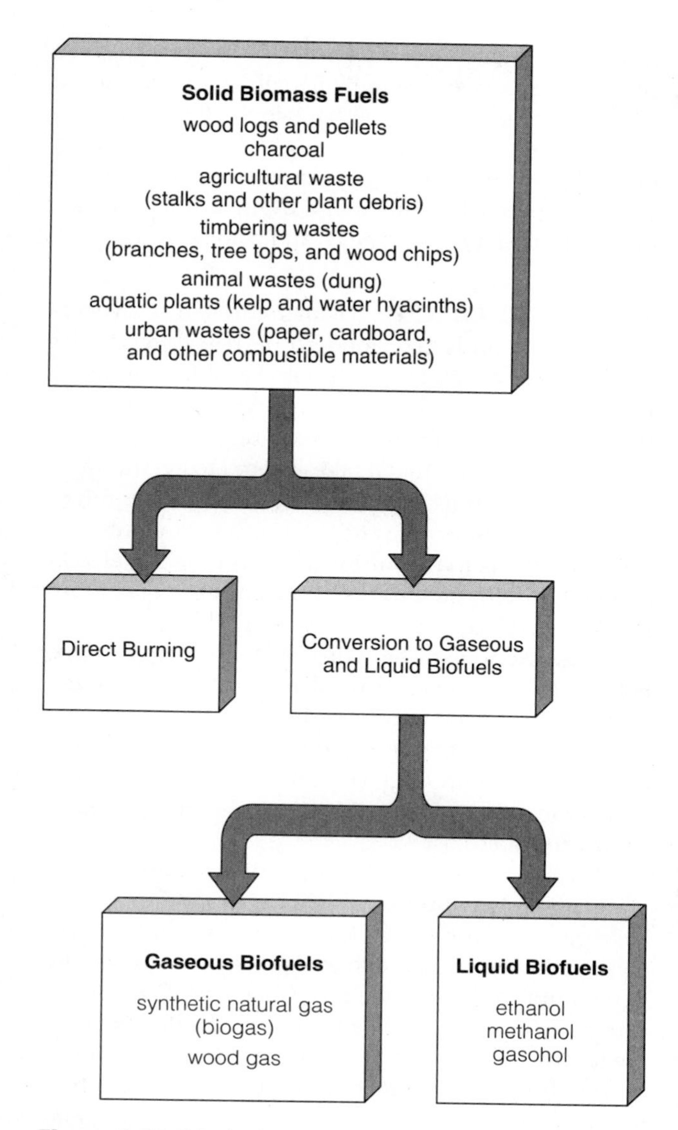

Figure 5-13 Principal types of biomass fuel.

cultural wastes (including animal manure), and some components of garbage. Some of this plant matter can be burned as solid fuel or converted into more convenient gaseous or liquid *biofuels* (Figure 5-13). Biomass—mostly from the burning of wood and manure to heat buildings and cook food—supplies about 11% of the world's energy (4–5% in Canada and the United States) and about 35% of the energy used in LDCs.

Various types of biomass fuels can be used for heating space and water, producing electricity, and propelling vehicles. Biomass is a renewable energy resource so long as trees and plants are not harvested faster than they grow back—a requirement that is not being met in many places (Section 4-3). Also, no net increase in atmospheric levels of heat-trapping carbon dioxide occurs so long as the rates of removal and burning of trees and plants—and the rate of loss of below-ground organic matter—do not exceed the rate of replenishment. Burning biomass fuels adds much less sulfur dioxide and nitric oxide to the atmosphere per unit of energy produced than does the uncontrolled burning of coal, and thus it requires fewer pollution controls.

However, it takes a lot of land to grow biomass fuel—about 10 times as much land as solar cells require to provide the same amount of electricity. Without effective land-use controls and replanting, widespread removal of trees and plants can deplete soil nutrients and cause excessive soil erosion, water pollution, flooding, and loss of wildlife habitat. Biomass resources also have a high moisture content (15–95%). The added weight of the moisture makes collecting and hauling wood and other plant material fairly expensive and reduces the net energy yield.

Sources of biomass include the following:

- *Biomass plantations*. In them, large numbers of fast-growing trees, shrubs, and water hyacinths are planted. After harvest, these "Btu bushes" can be burned directly, converted into burnable gas, or fermented into fuel alcohol. The plantations can be located on semiarid land not needed for growing crops (although lack of water can limit productivity) and can be planted to reduce soil erosion and help restore degraded lands. In some areas, plantations might compete with food crops for prime farm land. Conversion of large forested areas into single species biomass plantations also reduces biodiversity.
- *Wood and wood wastes*. Almost 70% of the people living in LDCs heat their dwellings and cook their food by burning wood or charcoal. However, at least 1.1 billion people in LDCs cannot find, or are too poor to buy, enough fuelwood to meet their needs, and that number may increase to 2.7 billion by 2000. Wood has a moderate to high net energy yield when collected and burned directly and efficiently near its source. However, in urban areas where wood must be hauled from long distances, it can cost homeowners more per unit of energy produced than oil or electricity. Harvesting wood can cause accidents (mostly from chain saws), and burning wood in poorly maintained or operated wood stoves can cause house fires. Wood stoves can also pollute the air, especially with particulate matter, causing some areas to ban the use of wood stoves. Since 1990 the EPA has required all new wood stoves sold in the United States to emit at least 70% less particulate matter than earlier models.
- *Agricultural wastes*. In agricultural areas, crop residues and animal manure can be collected and burned or converted into biofuels. This makes

A: \$4.2 billion (some say \$7 billion with more accurate bookkeeping) sense when residues or manure are burned in small power plants located near the site where the residues are produced. Otherwise, it takes too much energy to collect, dry, and transport the residues to power plants. Some ecologists argue that it makes more sense to use animal manure as a fertilizer and to use crop residues to feed livestock, retard soil erosion, and fertilize the soil.

- *Urban wastes.* An increasing number of cities in Japan, western Europe, and the United States have built incinerators that burn trash and use the energy released to produce electricity or to heat nearby buildings. However, this approach has been limited by opposition from citizens concerned about emissions of toxic gases and disposal of toxic ash. Some analysts argue that more energy is saved by composting or recycling paper and other organic wastes than by burning them.
- *Conversion of biomass into gaseous fuel.* In China, anaerobic bacteria in more than 6 million *biogas digesters* (500,000 of them improved models built in the 1980s) convert organic plant and animal wastes into methane fuel for heating and cooking. After the biogas has been separated, the solid residue is used as fertilizer on food crops or, if contaminated, on trees. When they work, biogas digesters are very efficient. However, they are slow and unpredictable, a problem that could be corrected with development of more reliable models. Methane gas can also be produced by anaerobic digestion of manure from animal feedlots and sludge from sewage treatment plants.
- *Conversion of biomass into liquid fuel.* Some analysts believe that liquid ethanol and methanol could replace gasoline and diesel fuel when oil becomes too scarce and expensive. Table 5-1 gives the advantages and disadvantages of using ethanol, methanol, and several other fuels as alternatives to gasoline.

5-5 The Solar-Hydrogen Revolution

GOODBYE OIL AND SMOG, HELLO HYDROGEN When oil is gone or what's left costs too much to use, what will fuel our vehicles, our industries, and our buildings? Some scientists say the fuel of the future is hydrogen gas (H_2).

There is very little hydrogen gas around, but we can get it from something we have plenty of: water, when split by electricity into gaseous hydrogen and oxygen (Figure 5-14, p. 96). If we can make the transition to an energy-efficient solar-hydrogen age, we could say goodbye to smog, oil spills, acid deposition, and nuclear energy, and perhaps to the threat of global warming. The reason is simple: When hydrogen burns in air it reacts with oxygen gas to produce water vapor (Figure 5-14)—not a bad thing to have coming out of our tailpipes, chimneys, and smokestacks.

WHAT'S THE CATCH? Using hydrogen as an energy source sounds too good to be true, doesn't it? You're right. We must solve several problems to make hydrogen one of our primary energy resources, and we're making rapid progress in doing this.

One problem is that it takes energy to get this marvelous fuel. The electricity needed to split water might come from coal-burning and nuclear power plants, but this subjects us to the harmful environmental effects associated with using these fuels, and it costs more than the hydrogen fuel is worth.

Most proponents of hydrogen as an energy source believe that the energy used to produce hydrogen from water must come from the sun. This means we must develop solar energy concentrators (Figure 5-9) or solar-cell technology (Figure 5-10) to the point at which the energy they produce can decompose water at an affordable cost.

If we can learn how to use sunlight to decompose water cheaply enough, we will set in motion a *solar-hydrogen revolution* over the next 50 years that will change the world as much as—if not more than—the agricultural and industrial revolutions did. Scientists—especially in Japan, Germany, and the United States—are hard at work trying to bring about this revolution.

Hydrogen gas is much easier to store than electricity. It can be stored in a pressurized tank or in metal powders that absorb gaseous hydrogen and release it when heated for use as a fuel in a car. If metal powder storage turns out to be economical, instead of pumping gas you might drive up to a fuel station, pull out a metal rack, replace it with a new one charged with metallic hydrogen, and zoom away. Unlike gasoline, solid metallic hydrogen compounds will not explode or burn if a vehicle's tank is ruptured in an accident. The problem is that it's difficult to store enough hydrogen gas in a car as a compressed gas or as a metal powder for it to run very far—a problem similar to the one current electric cars face. Scientists and engineers are seeking solutions to this problem.

We don't need to invent hydrogen-powered vehicles. Mercedes, BMW, and Mazda are already testing prototypes on the roads, and Mercedes hydrogen-powered buses could be on the streets of Hamburg, Germany, by 1997. And in Japan in 1992, Mazda unveiled a prototype car that runs on hydrogen released slowly from metal powder heated by the car's radiator coolant.

Q: How much of all U.S. land area is protected as wilderness?

Table 5-1 Evaluation of Alternatives to Gasoline

Advantages	Disadvantages
Compressed Natural Gas	
Fairly abundant domestic and global supplies Low emissions of hydrocarbons, CO, and CO_2 Currently inexpensive Vehicle development advanced Reduced engine maintenance Well suited for fleet vehicles Efficient	Cumbersome fuel tank required Expensive engine modification required ($2,000) One-fourth the range New filling stations required Nonrenewable resource
Electricity	
Renewable if not generated from fossil fuels or nuclear power Zero vehicle emissions Electric grid in place Efficient and quiet	Limited range and power Batteries expensive Slow refueling (6–8 hours) Power-plant emissions if generated from coal or oil
Reformulated Gasoline (Oxygenated Fuel)	
No new filling stations required Low to moderate reduction of CO emissions No engine modification required	Nonrenewable resource Dependence on imported oil perpetuated Possible high cost to modify refineries No reduction of CO_2 emissions Higher cost Water resources contaminated by leakage and spills
Methanol	
High octane Reduction of CO_2 emissions (unless produced by burning coal) Reduced total air pollution (30–40%)	Large fuel tank required One-half the range Corrosive to metal, rubber, plastic Increased emissions of potentially carcinogenic formaldehyde High CO_2 emissions if generated by coal High capital cost to produce Difficult to start in cold weather
Ethanol	
High octane Reduction of CO_2 emissions (total amount depends on distillation process and efficiency of crop growing) Reduction of CO emissions Potentially renewable	Large fuel tank required Much higher cost Corn supply limited Competition with food growing for cropland Lower range than gasoline Smog formation possible Corrosive Difficult to start in cold weather
Solar-Hydrogen	
Renewable if produced using solar energy Lower flammability Virtually emission-free Zero emissions of CO_2 Nontoxic	Nonrenewable if generated by fossil fuels or nuclear power Large fuel tank required No distribution system in place Engine redesign required Currently expensive

By 2000, hydrogen cars could be cost-competitive with gasoline cars in the United States, if gasoline prices rise to about 53¢ per liter ($2 per gallon)—half the current price in some European countries.

Another possibility is to power a car with a *fuel cell* in which hydrogen and oxygen gas combine to produce electrical current. Fuel cells have high energy efficiencies of up to 60%—several times the efficiency of conventional gasoline-powered engines and electric cars. Hydrogen-powered fuel cells may also be the best way to meet the heating and electricity needs of homes. Fuel cells are currently expensive, but this could change with more research and mass production. A fleet of 20 Mercedes-Benz hydrogen-powered test cars have logged over half-a-million miles powered by electricity from fuel cells.

A: 4% (only 1.8% in the lower 48 states)

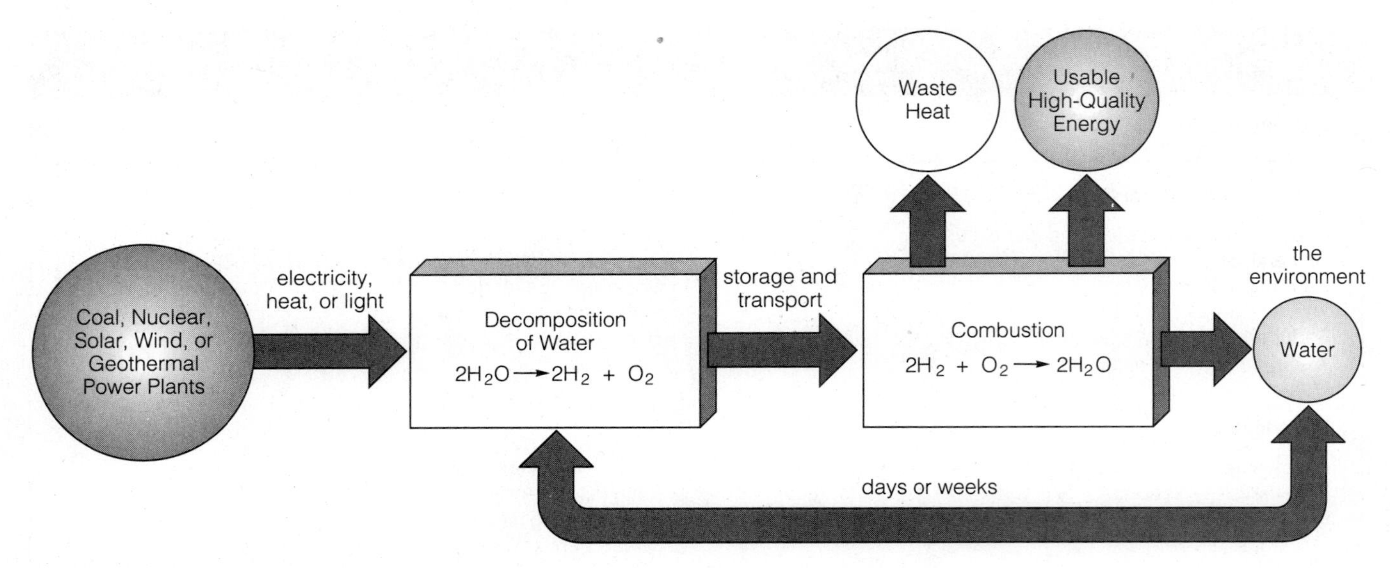

Figure 5-14 The hydrogen energy cycle. The production of hydrogen gas requires electricity, heat, or solar energy to decompose water, thus leading to a negative net energy yield. However, hydrogen is a clean-burning fuel that can be used to replace oil and other fossil fuels and nuclear energy. Using solar energy to produce hydrogen from water could also eliminate most air pollution and greatly reduce the threat of global warming.

Sometime in the 1990s a German firm plans to market solar-hydrogen systems that would meet all the heating, cooling, cooking, refrigeration, and electrical needs of a home, as well as providing hydrogen fuel for one or more cars.

WHAT'S HOLDING UP THE SOLAR-HYDROGEN REVOLUTION? Designing the technology for hydrogen-fueled cars, factories, home furnaces, and appliances is the easy part. The biggest problems are economic and political: figuring out how to replace an economy based largely on oil with entirely new production and distribution facilities and jobs. Overcoming these problems involves **(1)** convincing investors and energy companies with strong vested interests in fossil and nuclear fuels to risk lots of capital on hydrogen, and **(2)** convincing governments to put up some of the money for developing hydrogen energy, as they have done for decades for fossil fuels and nuclear energy. In the United States, large-scale government funding of hydrogen research is generally opposed by powerful U.S. oil companies, electric utilities, and automobile manufacturers, for a solar-hydrogen revolution represents a serious threat to their short-term economic well-being.

By contrast, the Japanese and German governments have been spending 7–8 times more on hydrogen research and development than the United States. Germany and Saudi Arabia have each built a large solar-hydrogen plant, and Germany and Russia have entered into an agreement for the joint development of hydrogen propulsion technology for commercial aircraft. Hydrogen has about 2.5 times the energy by weight of gasoline, making it an especially attractive aviation fuel. Without greatly increased private and government research and development, Americans will be buying solar-hydrogen equipment and fuel cells from Germany and Japan and will lose out on a huge global market and source of domestic jobs.

5-6 Geothermal Energy: Tapping Earth's Internal Heat

Heat contained in underground rocks and fluids is an important source of energy. Over millions of years this **geothermal energy** from the earth's mantle (Figure 1-3) has been transferred to underground concentrations of *dry steam* (steam with no water droplets), *wet steam* (a mixture of steam and water droplets), and *hot water* trapped in fractured or porous rock at various places in the earth's crust.

If geothermal sites are close to the surface, wells can be drilled to extract the dry steam, wet steam (Figure 5-1), or hot water. This thermal energy can be used for space heating and to produce electricity or high-temperature heat for industrial processes.

Geothermal reservoirs can be depleted if heat is removed faster than it is renewed by natural processes. Thus geothermal resources are nonrenewable on a human time scale, but the potential supply is so vast that it is often classified as a potentially renewable energy resource.

Q: How much of the world's population relies on plants or plant extracts for medicines?

Currently, about 20 countries are extracting energy from geothermal sites and are supplying enough heat both to meet the needs of over 2 million homes in a cold climate and to provide enough electricity for over 1.5 million homes. The United States accounts for 44% of the geothermal electricity generated worldwide, with most of the favorable sites in the West, especially in California and the Rocky Mountain states. Iceland, Japan, New Zealand, and Indonesia are among the countries with the greatest potential for tapping geothermal energy.

The biggest advantages of geothermal energy include a vast and sometimes renewable supply of energy for areas near geothermal sites; moderate net energy yields for large and easily accessible geothermal sites; far less carbon dioxide emission per unit of energy than fossil fuels; and costs that are competitive with other methods of producing electricity (Figure 5-7).

A serious limitation of geothermal energy is the scarcity of easily accessible geothermal sites. Geothermal reservoirs must also be carefully managed or they can be depleted within a few decades. Furthermore, geothermal development in some areas can destroy or degrade forests or other ecosystems. Without pollution control, geothermal energy production causes moderate to high air and water pollution. Noise, odor, and local climate changes can also be problems. With proper controls, however, most experts consider the environmental effects of geothermal energy to be no greater than, or less than, those of fossil-fuel and nuclear power plants.

5-7 Nonrenewable Fossil Fuels

CONVENTIONAL CRUDE OIL **Petroleum**, or **crude oil**, is a gooey liquid consisting mostly of hydrocarbon compounds, with small amounts of oxygen, sulfur, and nitrogen compounds. Crude oil and natural gas are often trapped together deep within the earth's crust (Figure 5-1), dispersed in pores and cracks in rock formations. A well can be drilled, and the crude oil that flows by gravity into the bottom of the well can be pumped out. Most crude oil travels by pipeline to a refinery, where it is heated and distilled to separate it into gasoline, heating oil, diesel oil, asphalt, and other components. Some of the resulting products, called *petrochemicals*, are used as raw materials in industrial chemicals, fertilizers, pesticides, plastics, synthetic fibers, paints, medicines, and many other products (Figure 5-4).

Oil is still cheap (when adjusted for inflation it costs about the same as it did in 1950) and easily transported within and between countries, and when extracted from easily accessible deposits it has a high net energy yield (Figure 5-3). Oil's low price has encouraged MDCs and LDCs alike to become heavily dependent on—indeed, addicted to—this important resource. Low prices have also encouraged wastage of oil and have discouraged the switch to other sources of energy and improvements in energy efficiency.

Oil *reserves* are identified deposits from which oil can be extracted profitably at current prices with current technology. The 13 countries that make up the Organization of Petroleum Exporting Countries (OPEC)* have 67% of these reserves. Saudi Arabia, with 25%, has the largest known crude oil reserves. Geologists believe that the politically volatile Middle East also contains most of the world's undiscovered oil. Therefore, OPEC is expected to have long-term control over world oil supplies and prices.

With only 4% of the world's oil reserves, the United States uses nearly 30% of the oil extracted worldwide each year, 63% of it for transportation. Despite an upsurge in exploration and test drilling, U.S. oil extraction has declined since 1985. As a result, the United States imported 51% of the oil it used in 1993, and by 2010 it could be importing 70%. This dependence, and the likelihood of much higher oil prices, could drain the United States (and other major oil-importing nations) of vast amounts of money, leading to severe inflation and widespread economic recession, perhaps even a major depression.

Oil's fatal flaw is that affordable supplies may be depleted within 35–80 years, depending on how rapidly it is used. At the current rate of consumption, known world oil reserves will last for 42 years. Undiscovered oil that is thought to exist might last another 20–40 years. At today's consumption rate, U.S. reserves will be depleted by 2018, and by 2010 if usage rises by 2% per year.

Some analysts argue that rising oil prices will stimulate exploration and that the earth's crust may contain 100 times more oil than is generally thought. Most geologists, however, do not believe that this oil exists. Some additional oil (less useful heavy oil) can be obtained from nonconventional sources, such as deposits of *oil shale* and *tar sand*. However, these deposits are not expected to be major sources of oil (except in a few places with concentrated supplies) because of low net energy yields, land degradation, pollution, and high costs.

*OPEC was formed in 1960 so that LDCs with much of the world's known and projected oil supplies could get a higher price for this resource. Today, its 13 members are Algeria, Ecuador, Gabon, Indonesia, Iran, Iraq, Kuwait, Libya, Nigeria, Qatar, Saudi Arabia, United Arab Emirates, and Venezuela.

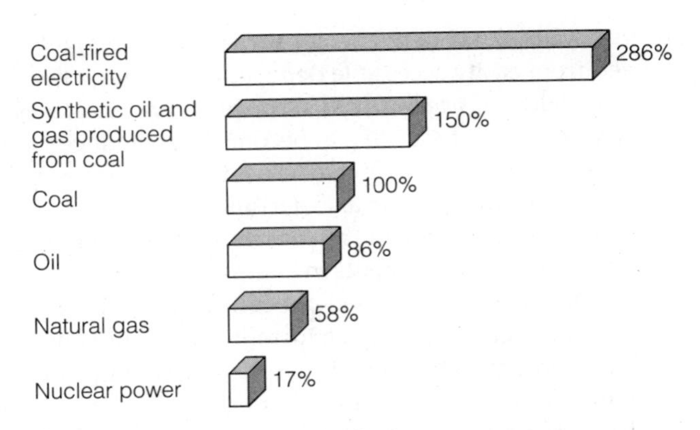

Figure 5-15 Carbon dioxide emissions per unit of energy produced by various fuels, expressed as percentages of emissions produced by coal.

Other analysts argue that optimistic projections about future oil supplies ignore the consequences of exponentially increasing consumption of oil. Just to continue using oil at the current rate and not run out, we must discover the equivalent of a new Saudi Arabian supply *every 10 years*. The estimated reserves under Alaska's North Slope—the largest oil field ever found in North America—would meet current world demand for only 6 months, or U.S. demand for 3 years. Moreover, if the economies of LDCs grow, the rate of oil consumption will probably exceed current rates and deplete supplies even faster.

Another drawback of oil (and all fossil fuels) is that burning it releases heat-trapping carbon dioxide (Figure 5-15), which could alter global climate (Section 6-2), as well as other air pollutants that damage people, crops, trees, fish, and other species (Section 6-4). Oil extraction and oil spills and leakage pollute water. If the harmful environmental effects of extracting, processing, and using oil (Figure 1-9) were included in its market price—and if current government subsidies keeping its price artificially low were removed—oil would be too expensive to use and would be replaced by a variety of less harmful and cheaper renewable energy resources.

NATURAL GAS In its underground gaseous state, **natural gas** is a mixture of 50–90% by volume of methane (CH_4) and smaller amounts of heavier gaseous hydrocarbons such as propane and butane. *Conventional natural gas* lies above most reservoirs of crude oil (Figure 5-1); *unconventional natural gas* is found by itself in several other underground sources. It is not yet economically feasible to get natural gas from unconventional sources, but the extraction technology is being developed rapidly.

When a natural-gas field is tapped, propane and butane gases are liquefied and removed as *liquefied petroleum gas (LPG)*. LPG is stored in pressurized tanks for use mostly in rural areas not served by natural-gas pipelines. The rest of the gas (mostly methane) is dried to remove water vapor, cleaned of hydrogen sulfide and other impurities, and pumped into pressurized pipelines for distribution. At a very low temperature of –184°C (–300°F), natural gas can be converted to **liquefied natural gas (LNG)**. This highly flammable liquid can then be shipped to other countries in refrigerated tanker ships.

The countries making up the former Soviet Union, with 40% of the world's natural gas reserves, are the world's largest extractors of natural gas. Other countries with large proven natural gas reserves are Iran (14%), the United States (5%), Qatar (4%), Algeria (4%), Saudi Arabia (3%), and Nigeria (3%). Geologists expect to find more natural gas, especially in LDCs.

So far, natural gas has been cheaper than oil. Known reserves and undiscovered, economically recoverable deposits of conventional natural gas in the United States are projected to last 60 years, and world supplies at least 80 years, at current consumption rates. It is estimated that conventional supplies of natural gas, as well as unconventional supplies available at higher prices, will last about 200 years at the current consumption rate, and 80 years if usage rates rise by 2% per year. Natural gas can be transported easily over land by pipeline, has a high net energy yield, and burns hotter and produces less air pollution than any other fossil fuel. Burning natural gas produces just over half as much heat-trapping carbon dioxide per unit of energy produced as coal, and two-thirds that of oil (Figure 5-15). And extracting natural gas damages the environment much less than extracting either coal or uranium ore.

Because of its advantages over oil and coal, some analysts see natural gas as the best fuel to help us make the transition to improved energy efficiency and much greater use of renewable energy sources over the next 50 years. Also, hydrogen gas produced from water by solar-generated electricity could be mixed gradually with natural gas to help smooth the shift to a solar-hydrogen energy system (Section 5-5).

COAL **Coal** is a solid formed in several stages as plant remains are subjected to intense heat and pressure over millions of years. It is used to generate 44% of the world's electricity (56% in the United States) and to make 75% of its steel. About 68% of the world's proven coal reserves and 85% of the estimated undiscovered coal deposits are located in the United States, the former Soviet Union, and China (which gets 76% of its commercial energy from coal).

Coal is the most abundant conventional fossil fuel in the United States and worldwide. Identified world reserves of coal should last about 220 years at current

Q: What percentage of the world's estimated plant species have been evaluated for their medical uses?

usage rates, and 65 years if usage rises by 2% per year. The world's unidentified coal reserves are projected to last about 900 years at the current consumption rate, and 149 years if the usage rate increases by 2% per year. Identified U.S. coal reserves should last about 300 years at the current consumption rate; unidentified U.S. coal resources could extend those supplies for perhaps 100 years, at a much higher average cost. Coal also has a high net energy yield (Figure 5-3).

However, coal has a number of drawbacks. Coal mining is dangerous because of accidents and black lung disease, a form of emphysema caused by prolonged inhalation of coal dust and other particulate matter. Underground mining causes land to sink when a mine shaft collapses during or after mining. Most surface-mined coal is removed by area strip mining or contour strip mining, depending on the terrain (Figure 5-1), and causes severe land disturbance and soil erosion. Restoration (if required) is expensive, and in arid and semiarid areas the land cannot be fully restored. Surface and subsurface coal mining can also severely pollute nearby streams and groundwater with acids and toxic metal compounds. Furthermore, once coal is mined it is expensive to move from one place to another, and it cannot be used in solid form as a fuel for cars and trucks.

Because coal produces more carbon dioxide per unit of energy than do other fossil fuels (Figure 5-15), burning more coal accelerates projected global warming (Section 6-2). Without expensive air-pollution control devices, burning coal produces more air pollution per unit of energy than any other fossil fuel. Each year air pollutants from coal burning kill thousands of people (with estimates ranging from 5,000 to 200,000) and cause at least 50,000 cases of respiratory disease and several billion dollars in property damage in the United States alone. However, new ways, such as fluidized-bed combustion, have been developed to burn coal more cleanly and efficiently.

Coal can also be converted into synthetic natural gas (SNG) by *coal gasification*, or into a liquid fuel such as methanol or synthetic gasoline by *coal liquefaction*. These synfuels can be transported by pipeline, and when burned they produce much less air pollution (except for carbon dioxide) than does solid coal. However, most analysts expect synfuels to play only a minor role as an energy resource in the next 30–50 years because of low net energy yields, high costs, and higher carbon dioxide emissions than coal (Figure 5-15).

If all of coal's harmful environmental costs (Figure 1-9) were included in its market price, and if government subsidies were removed, coal would become so expensive that it would likely be replaced by cheaper and less environmentally harmful, renewable energy resources.

5-8 Nonrenewable Nuclear Energy

NUCLEAR POWER: A FADING DREAM How did we get into nuclear power plants? The answer is a mixture of technological ingenuity and guilt. After the United States dropped atomic bombs on Hiroshima and Nagasaki, killing over 200,000 people and ending World War II, the scientists who developed the bomb and the elected officials responsible for its use were determined to show that the peaceful uses of atomic energy could outweigh the immense harm it had done. One part of this "Atoms for Peace" program was to use nuclear power to produce electricity.

Initially, U.S. utility companies were skeptical. However, they began developing nuclear power plants in the late 1950s because the Atomic Energy Commission (which had the conflicting roles of promoting and regulating nuclear power) promised them that nuclear power would produce electricity at a much lower cost than coal and other alternatives, and because the government paid about a quarter of the cost of constructing the first group of commercial reactors. Also, after insurance companies refused to cover more than a small part of the possible damages from a nuclear power-plant accident, Congress passed the Price-Anderson Act, which protected the nuclear industry and utilities from significant liability to the general public in case of accidents. It was an offer utility company officials could not resist.

In the 1950s, researchers predicted that by the end of the century 1,800 nuclear power plants would supply 21% of the world's commercial energy and 25% of that used in the United States. By 1993, after over 40 years of development and huge government subsidies, only 423 commercial nuclear reactors in 25 countries were producing only 17% of the world's electricity and less than 5% of its commercial energy.

In western Europe, plans to build more new nuclear power plants have come to a halt, except in France, where the government builds standardized plants and has discouraged public criticism of nuclear power. No new nuclear power plants have been ordered in the United States since 1978, and 119 previous orders have been canceled. In 1993, the 108 licensed commercial nuclear power plants in the United States generated about 20% of the country's electricity, a percentage that is expected to fall over the next two decades when many of the current reactors reach the end of their useful life.

What has happened to nuclear power? The answer is billion-dollar construction cost overruns, high operating costs, frequent malfunctions, false

A: 2%

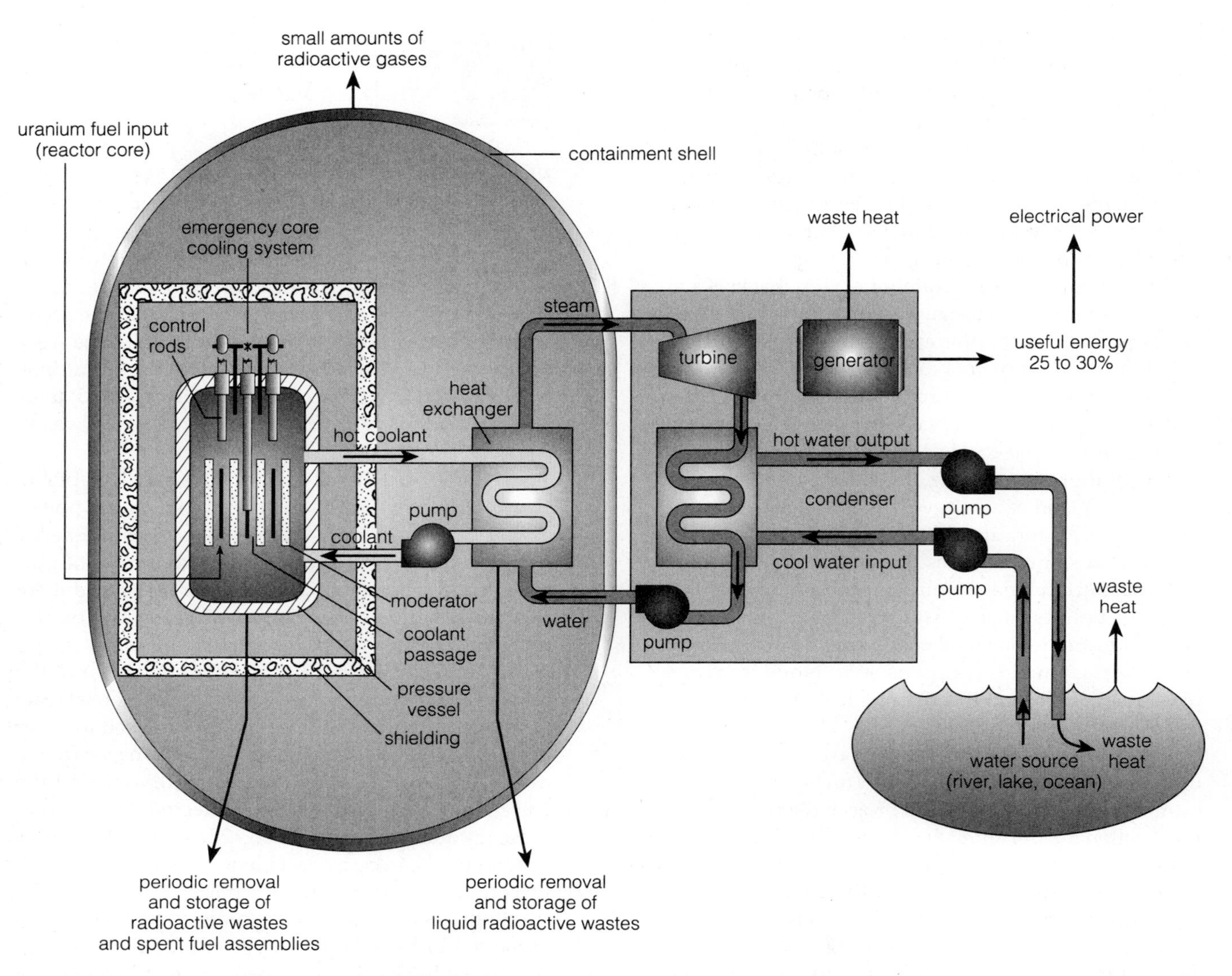

Figure 5-16 Light-water-moderated-and-cooled nuclear power plant with a pressurized water reactor. Worldwide, this type of plant generates about 85% of the electricity (100% in the United States) produced by nuclear power plants. The ill-fated Chernobyl nuclear reactor in Ukraine (a part of the former Soviet Union) was a carbon-moderated reactor with fewer extensive safety features than most reactors in the rest of the world.

assurances and cover-ups by government and industry officials, inflated estimates of electricity use, poor management, and public concerns about safety, costs, and radioactive waste disposal. To evaluate nuclear power, we need to know where nuclear energy comes from and how a nuclear power plant works.

NUCLEAR FISSION AND RADIOACTIVITY The source of energy for nuclear power is **nuclear fission**—a nuclear change in which nuclei of certain atoms (such as a type of uranium called uranium-235) are split apart into lighter nuclei when struck by neutrons; each fission releases two or three more neutrons and energy. Multiple fissions within a critical mass of the uranium fuel release an enormous amount of energy. The rate at which this happens can be controlled in the nuclear fission reactor of a nuclear power plant, and the heat generated can be used to spin a turbine and produce electricity (Figure 5-16).

Nuclear fission produces radioactive fission fragments containing unstable nuclei that spontaneously shoot out fast-moving particles, high-energy radiation (gamma rays), or both at a fixed rate. These unstable nuclei are called **radioactive isotopes**, or **radioisotopes**. This spontaneous process is called *radioactive decay*, and the particles and high-energy radiation released by radioisotopes are called *radioactivity*.

Radioactivity can cause changes or mutations in the DNA molecules that carry genetic information; harmful mutations can cause genetic defects in offspring or several generations later. Radioactivity can also damage living tissue, causing burns, miscarriages,

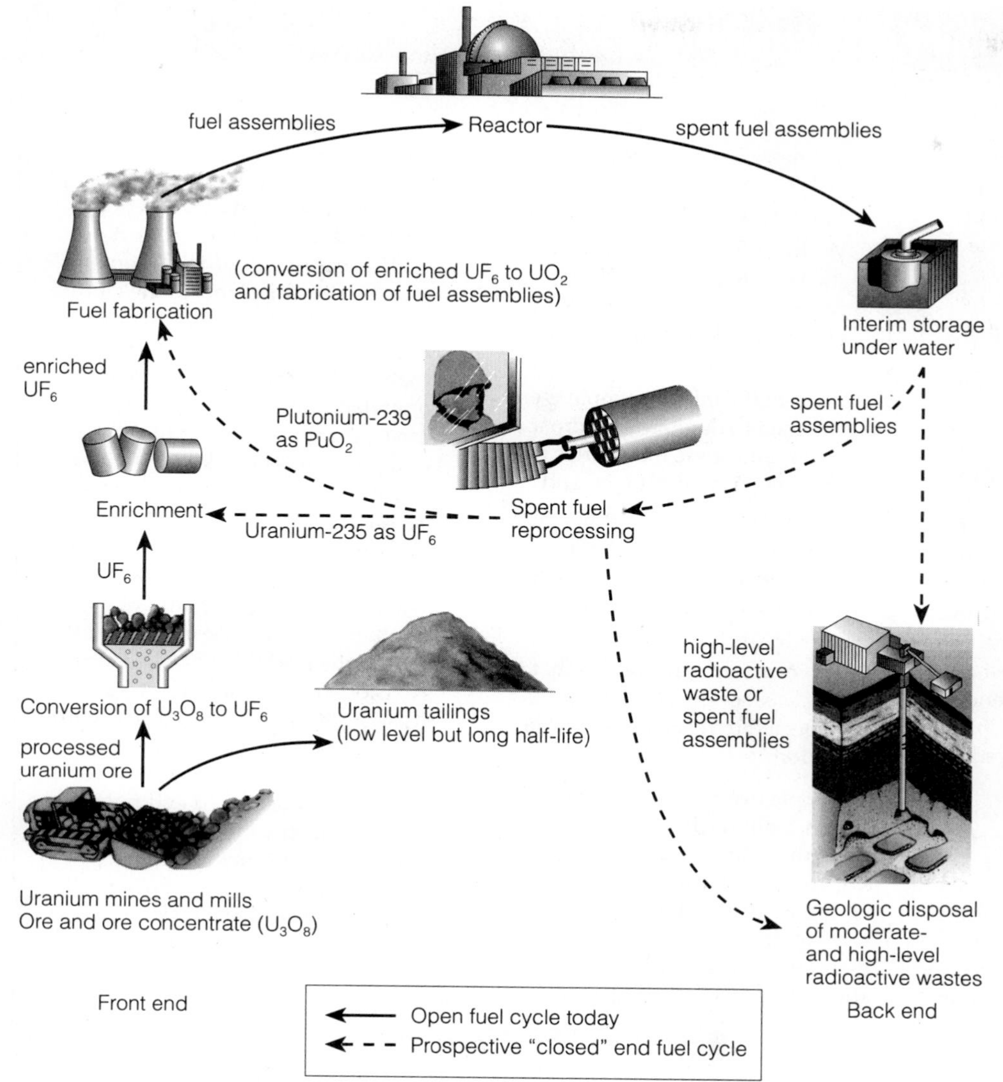

Figure 5-17 The nuclear fuel cycle.

cataracts, and cancers (bone, thyroid, breast, skin, and lung).

Nuclear power plants, each with one or more reactors, are only one part of the nuclear fuel cycle (Figure 5-17). In evaluating the safety and economy of nuclear power, we need to look at the entire cycle, not just the nuclear plant itself.

ADVANTAGES OF NUCLEAR POWER Nuclear power has some important advantages, including:

- *Nuclear plants don't emit air pollutants* (as do coal-fired plants) *so long as they are operating properly.*
- *The entire fuel cycle needed to mine uranium ore, convert it to nuclear fuel, run nuclear plants, and deal with nuclear wastes adds about one-sixth as much heat-trapping carbon dioxide per unit of electricity as does using coal,* thus making it more attractive than fossil fuels for reducing the threat of global warming.
- *Water pollution and disruption of land are low to moderate* if the entire nuclear fuel cycle operates normally.
- *Multiple safety systems with backups greatly decrease the likelihood of a catastrophic accident releasing deadly radioactive material into the environment.*

However, a partial or complete meltdown or explosion is possible, as the poorly designed and operated Chernobyl nuclear power plant in the former Soviet Union (now Ukraine) demonstrated (Connections, p. 102).

In 1979, one of the two reactors at the Three Mile Island (TMI) nuclear plant near Harrisburg, Pennsylvania, lost its coolant water because of a series of mechanical failures and human operator errors; about half of its core melted and fell to the bottom of the reactor. Investigators discovered that if a valve had stayed stuck open for another 30–60 minutes, there would have been a complete meltdown.

A: We don't know; estimates range from 5 million to 100 million

Bitter Lessons from Chernobyl

CONNECTIONS

Chernobyl is a chilling word recognized around the globe as the site of the worst nuclear disaster ever. On April 25, 1986, a series of explosions in a nuclear power plant in Ukraine (then in the Soviet Union) blew the massive roof off the reactor building and flung radioactive debris high into the atmosphere. Over the next several days winds carried some of those radioactive materials over parts of the Soviet Union and over much of eastern and western Europe as far as 2,000 kilometers (1,250 miles) from the plant. The accident happened when engineers turned off most of the reactor's automatic safety and warning systems to keep them from interfering with an unauthorized safety experiment.

The aftermath of the Chernobyl disaster includes the following consequences:

- 31 people died shortly after the accident from massive radiation exposure, and 259 people were hospitalized with acute radiation sickness. Some government officials have estimated the death toll to be more than 8,000 people.
- Within a few days, 135,000 people were evacuated; 125,000 more were evacuated in 1991, and 2.2 million more may need to be moved.
- Over half a million people were exposed to dangerous radioactivity, and some may suffer from cancers, thyroid tumors, and cataracts.
- 4 million people, mostly in Ukraine, Belarus, and northern Europe, may suffer health effects.
- Government officials say that the total cost of the accident is expected to reach at least $358 billion.

People evacuated from the region around Chernobyl had to leave their possessions behind and say goodbye, with little or no notice, to lush green wheat fields and blossoming apple trees, to land their families had farmed for generations, to cows and goats that would be shot because the grass they ate was radioactive, and to their contaminated cats and dogs. They will not be able to return.

Nuclear power officials contend that the nuclear plants in the United States, and most of those in other MDCs, have much safer designs than the Chernobyl reactor and are operated more safely, pointing to the fact that no member of the general public has been killed in these countries as a result of a nuclear accident.

Scientists in Germany and Sweden, however, project that worldwide there is a 70% chance of another serious nuclear accident within the next five years. Especially risky are 13 nuclear reactors in Russia, Lithuania, and Belarus with Chernobyl-type designs. Environmentalists urge that these and 10 other poorly designed and poorly operated nuclear plants in eastern Europe be shut down.

The primary lesson of Chernobyl is that *a major nuclear accident anywhere is a nuclear accident everywhere.*

No one can be shown to have died immediately because of the accident. Opponents of nuclear power point to the Three Mile Island accident as a reason to be concerned about the safety of nuclear power; supporters of nuclear power, however, consider the incident ample evidence that the built-in safety systems work, even though those systems could be (and since have been) improved.

PROBLEMS WITH NUCLEAR POWER Increased use of nuclear power has come to a virtual standstill in most countries, and its use will probably decrease as existing plants wear out faster than new ones are built. Problems connected with nuclear power include the following:

- *Long-term storage of radioactive wastes.* Each part of the nuclear fuel cycle (Figure 5-17) produces low-level and high-level solid, liquid, and gaseous radioactive wastes. The low-level wastes produced must be stored safely for several decades, and the high-level wastes for thousands of years. After 39 years of research and debate, scientists still don't agree on a safe method of storing these wastes. The most favored approach is sealing the material in metallic containers to be buried in presumably stable geological formations that are earthquake-resistant and waterproof (Figure 5-17). However, according to a 1990 report by the National Academy of Sciences, "use of geological information—to pretend to be able to make very accurate predictions of long-term site behavior—is scientifically unsound." Even if the problem is technically solvable, its solution may be politically unacceptable because most citizens oppose both storage of any radioactive wastes near their communities and the transport of such wastes through their communities on the way to storage sites.
- *What to do with worn-out nuclear plants.* After about 30–40 years a nuclear power plant ceases to

Q: What percentage of all species that have ever lived have become extinct?

SPOTLIGHT

Evaluating Nuclear Industry Advertisements

Since the Three Mile Island accident, the U.S. nuclear industry and utility companies have financed a $21-million-a-year public relations campaign by the U.S. Council for Energy Awareness. The campaign's goals are to improve the industry's image, resell nuclear power to the U.S. public, and minimize the importance of solar energy, energy efficiency, geothermal energy, wind, and hydropower as alternatives.

Most of the council's ads use the argument that the United States needs more nuclear power to reduce dependence on imported oil and improve national security. Sounds good, doesn't it? But since 1979 only about 5% (3% in 1991) of the electricity in the United States has been produced by burning oil, and 95% of that is residual oil that can't be used for other purposes. Even if all electricity in the United States came from nuclear power—which would require about 500 nuclear plants—U.S. oil consumption would be reduced by less than 5%. The nuclear industry also fails to point out that half of the uranium used for nuclear fuel in the United States is imported.

The U.S. nuclear industry hopes to persuade governments and utility companies to build hundreds of new "second-generation" plants using standardized designs with passive "fail-safe" features, which they claim are safer and can be built more quickly (in 3 to 5 years). However, according to *Nucleonics Week*, an important nuclear industry publication, "experts are flatly unconvinced that safety has been achieved—or even substantially increased—by the new designs." Furthermore, none of the new designs solves the problems of what to do with nuclear waste and worn-out nuclear plants, and of using nuclear technology and fuel to build nuclear weapons.

be safe to operate, but many of its parts will remain highly radioactive for thousands of years. The options are **(1)** dismantling the plant, **(2)** mothballing it for 30–100 years—by putting up a barrier and setting up a 24-hour-a-day security system—and then dismantling it, or **(3)** covering it with reinforced concrete and putting up a barrier to keep out intruders for several thousand years. Each method involves shutting down the plant, removing the spent radioactive fuel, and sending all radioactive materials to an approved waste storage site yet to be built. By 2030 all U.S. reactors will have to be retired, based on the life of their operating licenses, but many may be retired early because of safety problems or because they cost too much to run. Utility company officials estimate that retiring a typical large reactor should cost about $200 million, but most analysts place the cost at around $1 billion per reactor, adding to the already high cost of electricity produced by nuclear fission (Figure 5-7).

- *Use of reactors to make bomb-grade material for nuclear weapons*. We live in a world with enough nuclear weapons to kill everyone on Earth 60 times over—20 times over if current nuclear arms reduction agreements are carried out. By the end of this century 60 countries—one of every three in the world—will have either nuclear weapons or the knowledge and capability needed to build them. The bomb-grade fuel and knowledge required to build such nuclear weapons has come mostly from research conducted by, and commercial nuclear reactors sold by, the United States and 14 other countries since 1958. The planned dismantling of thousands of nuclear warheads in the United States and the former Soviet Union can increase this threat by leaving huge amounts of bomb-grade plutonium to be safeguarded. There is no truly effective solution to the serious problem of nuclear weapon proliferation.
- *The high cost of nuclear power*. Despite massive subsidies, the most modern nuclear power plants built in the United States produce electricity at a cost equivalent to electricity produced by burning oil costing well over $100 per barrel (compared to its current price of $18–20 per barrel). All methods of producing electricity in the United States (except solar photovoltaic and solar thermal plants) have average costs below those of new nuclear power plants (Figure 5-7). By 2000, even these methods (with relatively few subsidies) are expected to be cheaper than nuclear power. Banks and other lending institutions have become leery of financing new U.S. nuclear power plants. The Three Mile Island accident showed that utility companies could lose $1 billion worth of equipment in an hour and at least $1 billion more in cleanup costs, even without any established harmful effects on public health. Abandoned reactor projects have cost U.S. utility investors over $100 billion since the mid-1970s. *Forbes* magazine called the failure of the U.S. nuclear power program "the largest managerial disaster in U.S. business history." Despite massive economic and public relations setbacks, the U.S. nuclear power industry hopes for a comeback (Spotlight, left).

BREEDER NUCLEAR FISSION Some nuclear power proponents urge the development and widespread use of *breeder nuclear fission reactors*, which generate more nuclear fuel than they consume by converting a nonfissionable form of uranium (uranium-238) into fissionable nuclei of plutonium (plutonium-239). Because breeders would use over 99% of the uranium in ore deposits, the world's known uranium reserves would last at least for 1,000 years, and perhaps for several thousand years.

Since 1966, small experimental breeder reactors have been built in the United Kingdom, the former Soviet Union, Germany, Japan, and France. In December 1986 France opened a $3-billion commercial-size breeder reactor. Not only did it cost three times the original estimate to build, but the little electricity it has produced was twice as expensive as that generated by France's conventional fission reactors. In 1987, shortly after the reactor began operating at full power, it began leaking liquid sodium coolant and was shut down. Repairs may be so expensive that the reactor will never be put back into operation.

Tentative plans to build full-size commercial breeders in other countries have been abandoned because of the French experience and an excess of electric generating capacity and conventional uranium fuel. Also, the experimental breeder reactors already built produce plutonium fuel much too slowly. If this serious problem is not solved, it would take 100–200 years for breeders to begin producing enough plutonium to fuel a significant number of other breeder reactors.

NUCLEAR FUSION: FORCING NUCLEI TO COMBINE **Nuclear fusion** occurs when extremely high temperatures are used to force together the nuclei of light elements, such as hydrogen, until they fuse to form a heavier nucleus, releasing energy in the process. Temperatures of at least 100 million°C are needed to force the positively charged nuclei (which strongly repel one another) to fuse.

Despite almost 50 years of research and huge expenditures of mostly government funds, controlled nuclear fusion is still in the laboratory stage. So far, none of the approaches investigated has produced more energy than it uses. In 1989 two chemists claimed to have achieved nuclear fusion at room temperature using a simple apparatus, but subsequent experiments could not substantiate their claims.

If researchers can get more energy out of nuclear fusion than they put in, the next step would be to build a small fusion reactor and then scale it up to commercial size—one of the most difficult engineering problems ever undertaken. Also, the estimated cost of a commercial fusion reactor is several times that of a comparable conventional fission reactor.

If things go right, a commercial nuclear fusion power plant might be built by 2030. Even if everything goes right, however, energy experts don't expect nuclear fusion to be a significant energy source until 2100, if then.

5-9 Solutions: A Sustainable Energy Strategy

Many scientists and energy experts who have evaluated the energy alternatives discussed in this chapter have come to the following general conclusions:

- *The best short-term, intermediate, and long-term alternatives are a combination of improved energy efficiency and greatly increased use of locally available renewable energy resources.*
- *Future energy alternatives will probably have low-to-moderate net energy yields and moderate-to-high development costs.*
- *Because there is not enough financial capital to develop all energy alternatives, projects must be chosen carefully.*
- *We cannot and should not depend mostly on a single nonrenewable energy resource such as oil, coal, natural gas, or nuclear power.*

There are leaders in making the transition from the age of oil to the age of energy efficiency and renewable energy. Brazil and Norway get more than half their energy from hydropower, wood, and alcohol fuel. Israel, Japan, the Philippines, and Sweden plan to rely on renewable sources for most of their energy. California has become the world's showcase for solar and wind power. Each of us can help make the transition to a new age built around improved energy efficiency and increased use of renewable energy, and in the process we can save money and work with the earth (Individuals Matter, above right).

In the long run, humanity has no choice but to rely on renewable energy. No matter how abundant they seem today, eventually coal and uranium will run out.

DANIEL DEUDNEY AND CHRISTOPHER FLAVIN

Q: How many of the earth's species are believed to become extinct each year because of human activities?

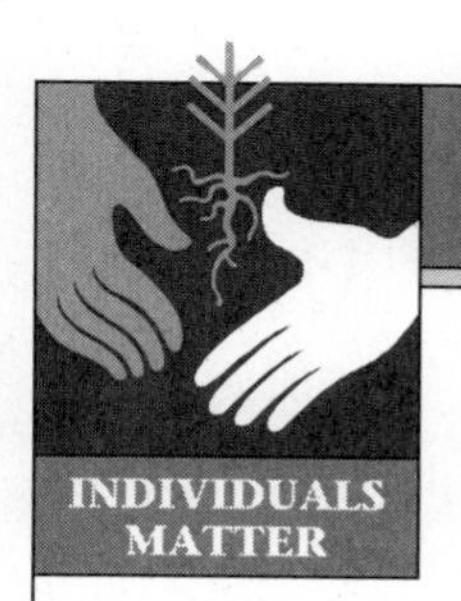

What *You* Can Do to Save Energy and Money

INDIVIDUALS MATTER

- *Don't use electricity to heat space or water* (unless it is provided by affordable solar cells or hydrogen).
- *Insulate new or existing houses heavily, caulk and weatherstrip to reduce air infiltration and heat loss, and use energy-efficient windows.* Add an air-to-air heat exchanger to minimize indoor air pollution.
- *Obtain as much heat and cooling as possible from natural sources*—especially sun, wind, geothermal energy, and trees (windbreaks and natural shading).
- *Buy the most energy-efficient home, lights, car, and appliances available,* and evaluate them in terms of lifetime cost.
- *Consider walking or riding a bicycle for short trips and buses or trains for long trips.*
- *Turn down the thermostat on water heaters to 43°–49° C (110°–120°F) and insulate hot water pipes.*
- *Lower the cooling load on air conditioners by raising the thermostat setting, installing energy-efficient lighting, using floor and ceiling fans, and using whole-house window or attic fans to bring in outside air (especially at night when temperatures are cooler).*
- *Turn off lights and appliances when not in use.*

CRITICAL THINKING

1. Suppose that a presidential candidate ran on a platform calling for the federal government to phase in a tax on gasoline over 5–10 years, so that eventually gasoline would cost about $3–5 a gallon (as is the case in Japan and most western-European nations). The candidate argues that this tax increase is necessary to encourage conservation of oil and gasoline, to reduce air pollution, and to enhance future economic, environmental, and military security. Some of the tax revenue would be used to provide tax relief or other aid to all people with incomes below a certain level (the poor and the lower-middle class), who would be hardest hit by such a consumption tax, and income taxes would be reduced to help compensate for the tax increase. Would you vote for this candidate who promises to triple the price of gasoline? Explain the merits of and problems with this candidate's position.

2. Explain why you agree or disagree with the following proposals by various analysts:
   a. The United States should cut average per capita energy use by at least 50% over the next 20 years.
   b. To solve world and U.S. energy supply problems, all we need do is recycle some or most of the energy we use.
   c. To solve present and future U.S. energy problems, all we need to do is find and develop more domestic supplies of oil, natural gas, and coal and increase dependence on nuclear power.
   d. The United States should institute a crash program to develop solar photovoltaic cells and solar-produced hydrogen fuel.
   e. Federal subsidies for all energy alternatives should be eliminated so that all energy choices can compete in a true free-market system.
   f. All government tax breaks and other subsidies for conventional fuels (oil, natural gas, coal), synthetic natural gas and oil, and nuclear power should be removed—and replaced with subsidies and tax breaks for improving energy efficiency and developing solar, wind, geothermal, and biomass energy alternatives.
   g. Development of solar energy alternatives should be left to private enterprise, with little or no help from the federal government, but nuclear energy and fossil fuels should continue to receive large federal subsidies.
   h. Between 2000 and 2020 the United States should phase out all nuclear power plants.
   i. The licensing time for new nuclear power plants in the United States should be halved (from an average of 12 years) so that they can be built at lower cost and can compete more effectively with coal-burning and other energy-producing facilities or technologies.
   j. A large number of new, better-designed nuclear fission power plants should be built in the United States to reduce dependence on imported oil and slow down projected global warming.

A: 4,000–36,000 (11 to 99 per day, but no one really knows how many)

6 The Atmosphere, Global Warming, and Ozone Loss

What's the use of a house if you don't have a decent planet to put it on?

Henry David Thoreau

6-1 The Atmosphere and Life on Earth

THE TROPOSPHERE: LIFE GIVER AND WEATHER BREEDER We live at the bottom of a "sea" of air called the **atmosphere** (Figure 1-3). About 75% of the mass of the earth's air is found in the atmosphere's innermost layer, the **troposphere**, which extends only about 17 kilometers (11 miles) above sea level at the equator and about 8 kilometers (5 miles) over the poles. If the earth were an apple, this lower layer, containing the air we breathe, would be no thicker than the apple's skin. This thin and turbulent layer of rising and falling air currents and winds is the planet's weather breeder.

The composition of the atmosphere has varied considerably throughout the earth's long history. Today about 99% of the volume of clean, dry air in the troposphere consists of two gases: nitrogen (78%) and oxygen (21%). The remaining 1% consists mostly of argon (Ar), carbon dioxide (CO_2), variable amounts of water vapor, and trace amounts of several other gases.

THE STRATOSPHERE: GLOBAL SUNSCREEN The atmosphere's second layer is the **stratosphere**, which extends from about 17–48 kilometers (11–30 miles) above the earth's surface. Although the stratosphere contains less matter than the troposphere, its composition is similar, with two notable exceptions: Its volume of water vapor is about 1,000 times less, and its volume of ozone (O_3) is about 1,000 times greater.

Ozone is produced in the stratosphere when some of the oxygen molecules there interact with lightning and solar radiation. Ozone is continuously being formed and destroyed, but as long as the rates of these two reversible processes are equal, the *average* concentration of ozone in the stratosphere remains constant (although the concentration varies at different altitudes and at different places). The presence of ozone in the stratosphere keeps about 99% of the sun's harmful ultraviolet (UV) radiation from reaching the earth's surface. This filtering action protects the human body from sunburn, skin and eye cancer, cataracts, and damage to the immune system. This "global sunscreen" also prevents damage to some plants, aquatic organisms, and other land animals. Furthermore, by keeping most UV radiation from reaching the troposphere, ozone in the stratosphere prevents much of the oxygen in the troposphere from being converted to toxic ozone, which is harmful to us and other forms of life. The trace amounts of ozone that do form in the troposphere as a component of urban smog damage plants, the respiratory systems of humans and other animals, and materials such as rubber.

Thus our good health, and that of many other species, depends on having enough "good" ozone in the stratosphere and as little "bad" ozone as possible in the troposphere. Unfortunately, our activities are both increasing the amount of harmful ozone in the tropospheric air we must breathe and decreasing the amount of beneficial ozone in the stratosphere.

HOW WEATHER AND CLIMATE AFFECT LIFE At every moment at any spot on Earth, the troposphere has a particular set of physical properties, including such things as temperature, pressure, humidity, precipitation, sunshine, cloud cover, and wind direction and speed. These short-term properties of the troposphere at a given place and time are what we call **weather**.

Climate is the long-term average weather of an area; it is a region's general pattern of weather conditions, seasonal variations, and weather extremes (such as hurricanes and prolonged drought or rain) over a long period (at least 30 years). The two most important factors determining the climate of an area are *temperature* and *precipitation*. Thus a land area's average amount of precipitation, together with its average temperature and soil type, generally determines whether

Q: What is the greatest threat to wild species?

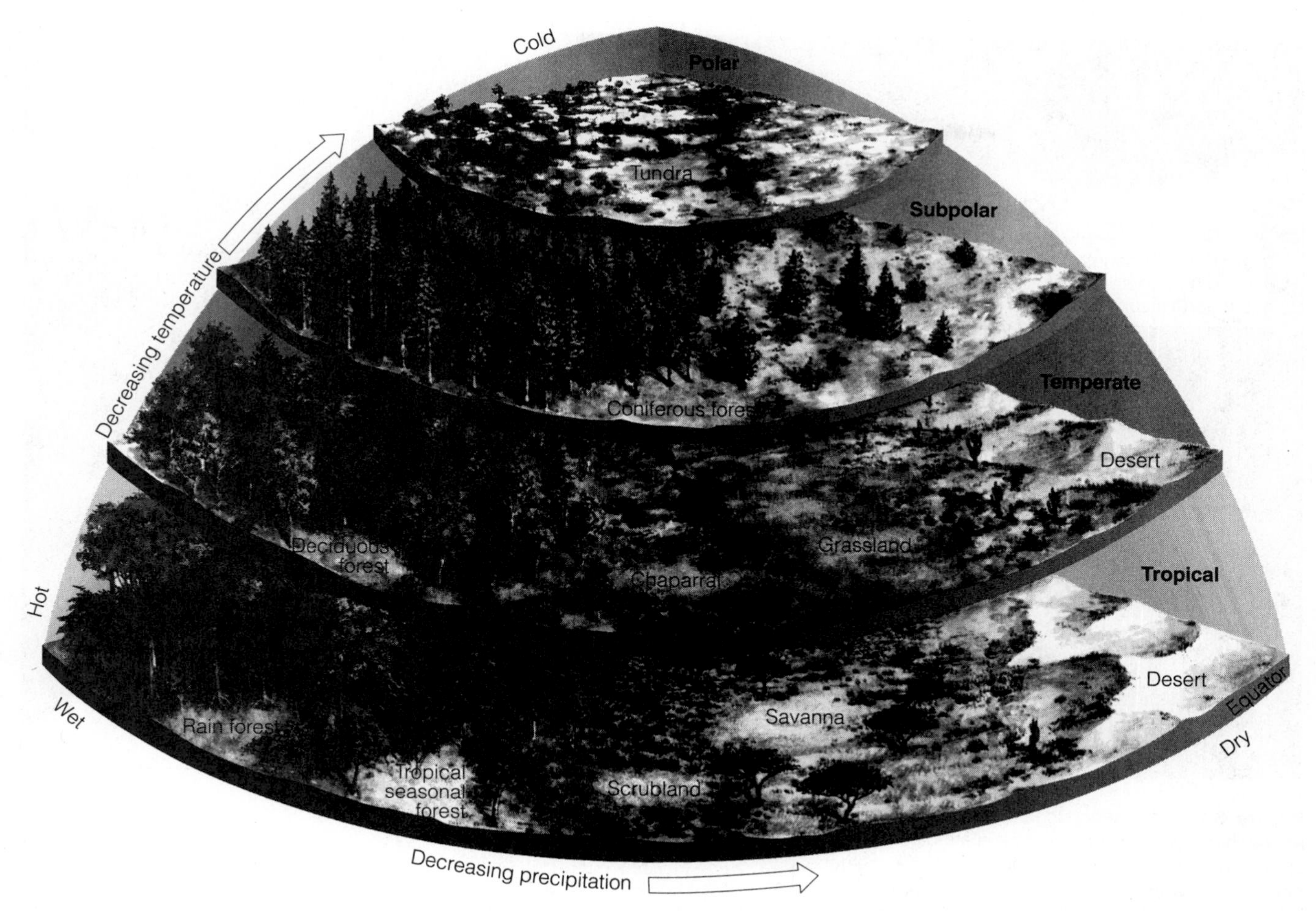

Figure 6-1 Average precipitation and average temperature, acting together over a period of 30 or more years as limiting factors, determine the type of desert, grassland, or forest found in a particular area. Although the actual situation is much more complex, this simplified diagram explains in a general way how climate determines the types and amounts of natural vegetation found in an area that has not been disturbed by human activities. (Used by permission of Macmillan Publishing Company from Derek Elsom, *Earth*, New York: Macmillan, 1992. Copyright 1992 by Marshall Editions Development Limited)

that area is desert, grassland, or forest. Acting together, these factors produce tropical, temperate, and polar deserts, grasslands, and forests (Figure 6-1).

THE SIGNIFICANCE OF THE GREENHOUSE EFFECT One of the many interacting factors affecting climate is the *chemical composition of the atmosphere.* Small amounts of carbon dioxide and water vapor, and trace amounts of ozone, methane, nitrous oxide, chlorofluorocarbons, and other gases in the troposphere, play a key role in determining the earth's average temperatures and thus its climates. These gases, known as **greenhouse gases**, act somewhat like the glass panes of a greenhouse (or of a car parked in the sun with its windows rolled up): They allow light, infrared radiation, and some ultraviolet radiation from the sun to pass through the troposphere. The earth's surface absorbs much of this solar energy and degrades it to heat, which then rises into the troposphere. Some of this heat escapes into space; some is absorbed by molecules of greenhouse gases, warming the air; and some radiates back toward the earth's surface. This trapping of heat in the troposphere is called the **greenhouse effect** (Figure 6-2). If there were no greenhouse gases, especially water vapor, the earth would be a cold and lifeless planet with an average surface temperature of –18°C (0°F) instead of its current 15°C (59°F) (Connections, p. 109).

Figure 6-2 The greenhouse effect. The rate at which heat flows from the earth, through the atmosphere, and back into space is affected by heat-trapping (greenhouse) gases in the troposphere, such as water vapor, carbon dioxide, methane, nitrous oxide, and ozone. Without the atmospheric warming provided by this natural effect, the earth would be a cold and mostly lifeless planet. (Used by permission from Cecie Starr and Ralph Taggart, *Biology: The Unity and Diversity of Life*, 6th ed., Belmont, Calif.: Wadsworth, 1992)

6-2 Global Warming? Or a Lot of Hot Air?

WHAT WE KNOW ABOUT THE EARTH'S CLIMATE The greenhouse effect is among the most widely tested and accepted scientific theories. However, there is much debate over whether human activities are now or will soon be intensifying the natural greenhouse effect, thereby raising the earth's average temperature (global warming) and changing the climate found in various parts of the world. There is also controversy over how much the temperature might rise and how this rise might affect the climate in different areas. If we are to evaluate the possibility of global warming and its possible effects, we need to look at what we know and don't know about the earth's climate.

In 1990 and 1992, the Intergovernmental Panel on Climate Change (IPCC) published reports by several hundred leading atmospheric scientists on the best available evidence concerning past climate change, the greenhouse effect, and recent changes in global temperatures. Based on the panel's reports and other studies, the following points constitute current scientific consensus on these matters:

- Earth's average surface temperature has fluctuated considerably over geologic time, including several ice ages that covered much of the planet with thick ice during the past 800,000 years. Each glacial period lasted about 100,000 years and was followed by a warmer interglacial period of 10,000–12,500 years. As the ice melted at the end of the last ice age, average sea levels rose about 100 meters (300 feet), greatly reducing the amount of dry land.

Q: How much of the earth's land surface has been set aside to protect wildlife?

- For the past 10,000 years we have enjoyed the relative warmth of the latest interglacial period, during which mean surface temperatures have usually fluctuated only 0.5–1°C (0.6–1.8°F) over 100- to 200-year periods. However, even these small temperature changes have led to large migrations of peoples in response to changed agricultural and grazing conditions.
- Over the past 160,000 years, levels of water vapor in the troposphere have remained fairly constant while those of carbon dioxide have fluctuated by a factor of two as part of the global carbon cycle (Figure 6-3). Estimated changes in the levels of tropospheric carbon dioxide (Figure 6-4) correlate fairly closely with estimated variations in the earth's mean surface temperature.
- Recent data from analysis of carbon dioxide and other gases trapped in layers of ice formed over the past 260,000 years reveal that the earth's climate has shifted often, drastically, and sometimes surprisingly quickly. The new evidence indicates that average temperatures during the warm interglacial period that began about 125,000 years ago varied as much as 10°C (18°F) in only a decade or two.
- Measured atmospheric levels of certain greenhouse gases—carbon dioxide, methane, nitrous oxide, and chlorofluorocarbons (CFCs)—have risen in recent decades.
- Most of the increased levels of these greenhouse gases have been caused by human activities: burning fossil fuels, use of chlorofluorocarbons, agriculture, and deforestation.
- Since 1860, when measurements began, mean global temperature has risen 0.3–0.6°C (0.5–1.1°F) (Figure 6-5).
- Eight of the 13 years from 1980 to 1992 were among the hottest in the 110-year recorded history of land-surface temperature measurements, and 1990 was the hottest of all.
- So far, any temperature changes possibly caused by an enhanced greenhouse effect have been too small to exceed normal short-term variations in mean atmospheric temperature caused by volcanic eruptions, air pollution, and other climatic factors.
- Warming or cooling by more than 2°C (4°F) over a few decades would be disastrous for the earth's ecosystems and for human economic and social systems. Such rapid climate change would alter conditions faster than some species, especially plants, could adapt or migrate. It might also shift areas where people could grow food. Some areas might become uninhabitable because of drought or because of floods following a rise in average sea levels.
- We don't know enough about how the earth works to make accurate projections about the possible effects of our inputs of greenhouse gases

CONNECTIONS

Earth: The Just-Right, Resilient Planet

Like Goldilocks tasting porridge at the house of the three bears, life as we know it is picky about temperature: Venus is much too hot, Mars is much too cold, and the earth is just right. Otherwise, you wouldn't be reading these words.

Life as we know it also depends on water. Again, temperature is crucial. Life on the earth needs average temperatures between the freezing and boiling points of water—between 0°C and 100°C (32°F and 212°F) at the earth's atmospheric pressures.

Earth's orbit is the right distance from the sun to have these conditions. If the earth were much closer, it would be too hot—like Venus—for water vapor to condense to form rain. If it were much farther, its surface would be so cold—like Mars—that its water would exist only as ice. The earth also spins; if it didn't, the side facing the sun would be too hot (and the other side too cold) for water-based life to exist. So far, the temperature has been, like Baby Bear's porridge, just right.

The earth is also the right size; that is, it has enough gravitational mass to keep its iron-nickel core molten and to keep its atmosphere from dissipating into space. The slow transfer of its internal heat (geothermal energy) to the surface also helps keep the planet at the right temperature for life. A much smaller Earth would not have enough gravitational mass to hold onto an atmosphere consisting of light molecules of gases such as nitrogen, oxygen, carbon dioxide, and water vapor. And thanks to the development of photosynthesizing bacteria over 2 billion years ago, an ozone sunscreen protects us from an overdose of ultraviolet radiation.

On a time scale of millions of years, the earth is also enormously resilient and adaptive. Its average temperatures have remained between the freezing and boiling points of water, even though the sun's energy output has increased by about 30% over the 3.6 billion years since life arose. In short, the earth is just right for life as we know it to have arisen.

A: 4.9%, compared to a minimum of 10% environmentalists say is needed

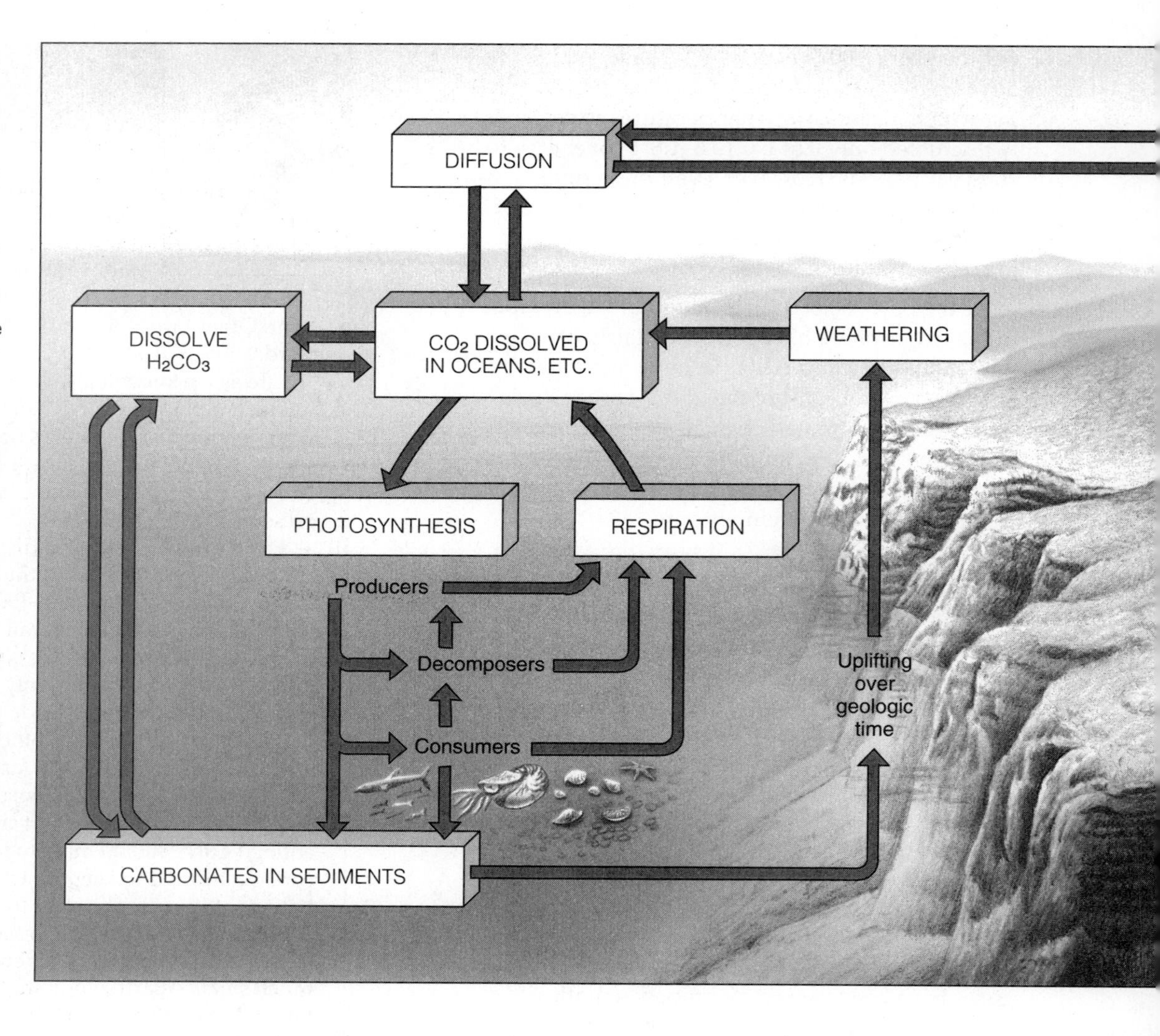

Figure 6-3 Simplified diagram of the global carbon cycle. The left portion shows the movement of carbon through marine ecosystems, and the right portion shows its movement through terrestrial ecosystems. If the carbon cycle removes too much CO_2 from the atmosphere, the earth will cool; if the cycle generates too much, the earth will get hotter. Especially since 1950, we have disturbed the carbon cycle in two ways that add more carbon dioxide to the atmosphere than oceans and plants have been able to remove: **(1)** forest and brush clearing, leaving less vegetation to absorb CO_2, and **(2)** burning fossil fuels and wood, which produces CO_2 that flows into the atmosphere. (Used by permission from Cecie Starr and Ralph Taggart, *Biology: The Unity and Diversity of Life*, 6th ed., Belmont, Calif.: Wadsworth, 1992)

either on global and regional climates or on the biosphere.

COMPUTERS AS CRYSTAL BALLS: MODELING GREENHOUSE WARMING To project the behavior of climatic and other complex systems, scientists develop mathematical models that simulate such systems and then run them on computers (Figures 2-9 and 2-10). How well the results correspond to the real world depends both on the design of the model and on the accuracy of the data and the assumptions used.

Current climate models generally agree on how global climate might change but disagree on changes for individual regions. Here are the main projections of major climate models:

- The earth's mean surface temperature will rise 1.5–5.5°C (2.7–9.9°F) by 2050 if inputs of greenhouse gases continue to rise at the current rate (Figure 6-5, p. 212). Even at the lower value, the earth would be warmer than it has been for 10,000 years.
- The Northern Hemisphere will warm more and faster than the Southern Hemisphere, mostly because the latter has more ocean (which can absorb more heat than land can).
- Temperatures at middle and high latitudes should rise two to three times the global average, whereas temperatures near the equator should rise less than the global average.
- More areas will have extreme heat waves and more forest and brush fires.
- The average sea level will rise 2–4 centimeters (0.8–1.6 inches) per decade.

A CLOUDY CRYSTAL BALL: WHAT WE DON'T KNOW Because we have only partial knowledge about how the earth's climate system works, our models and projections are flawed, but they are all we

Q: How much of the commercial energy used in the world comes from nonrenewable resources?

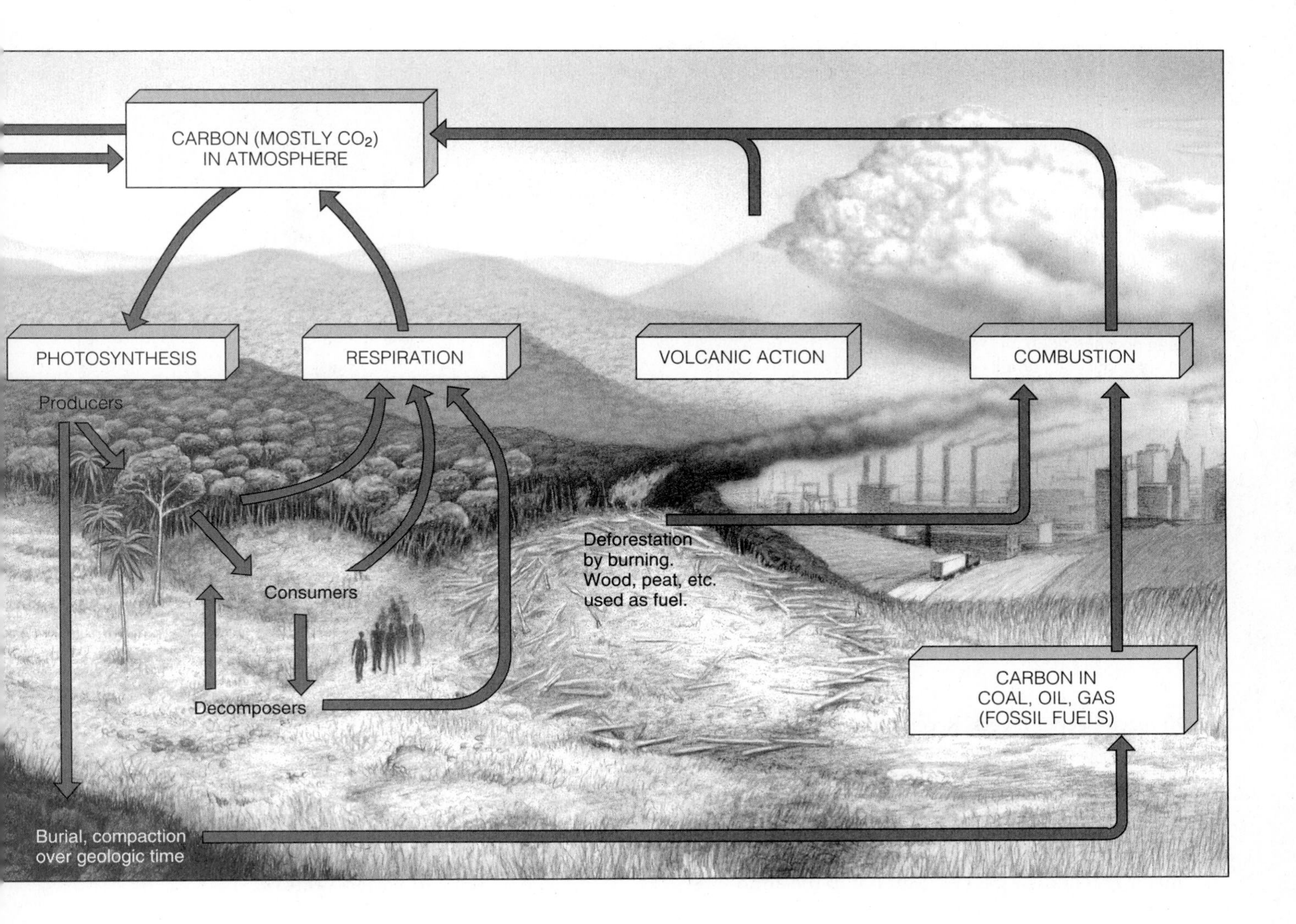

Figure 6-4 Estimated long-term variations in mean global surface temperature and average tropospheric carbon dioxide levels over the past 160,000 years. Since the last great ice age ended about 10,000 years ago, we have enjoyed a warm interglacial period. One factor in the earth's mean surface temperature is the greenhouse effect. Changes in tropospheric levels of carbon dioxide, a major greenhouse gas, correlate closely with changes in the earth's mean surface temperature and thus its climate, although other factors also influence global climate.

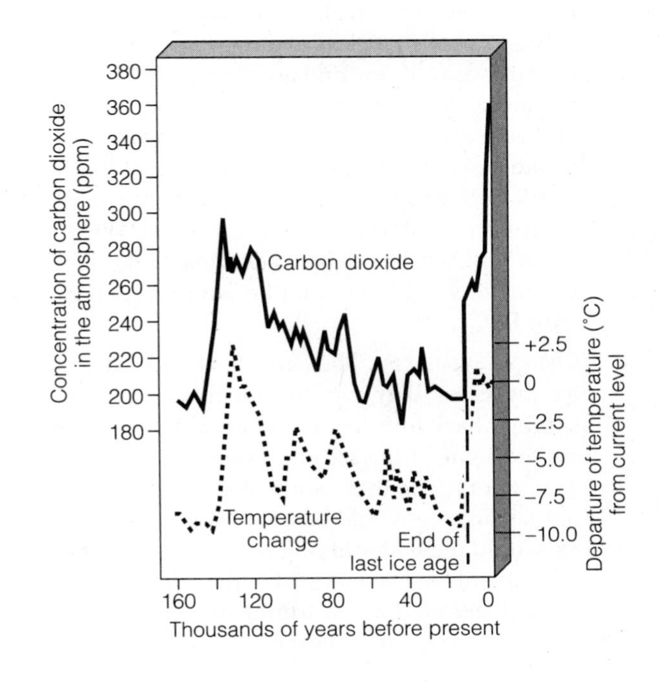

have. The following list contains some factors that might *dampen* or *amplify* a rise in average atmospheric temperature, determine how fast temperatures might climb or fall, and influence what the effects might be on various areas:

- *Changes in the amount of solar energy reaching the earth.*
- *Effects of oceans on climate.* The world's oceans could slow global warming by absorbing more heat, but this depends on how long the heat takes

A: 83% (78% from fossil fuels and 5% from nuclear power)

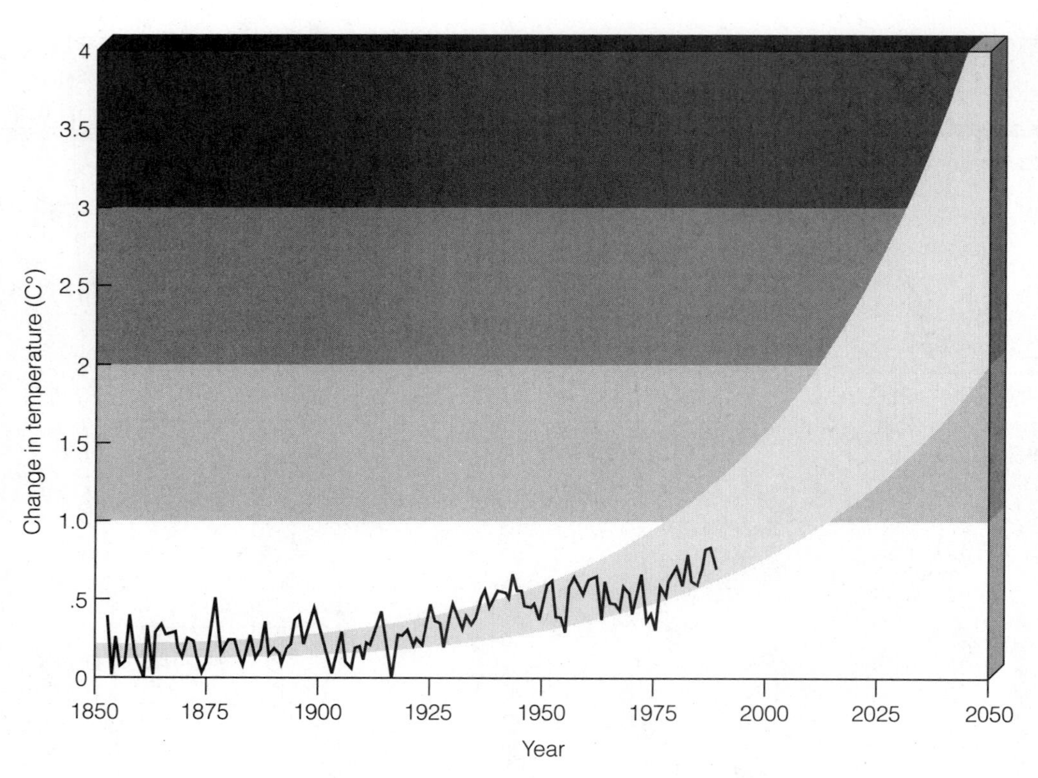

Figure 6-5 Changes in the earth's mean surface temperature between 1860 and 1990 (dark line). The upward-curving region shows global warming projected by various computer models of the earth's climate systems. Note that the computer projections *roughly* match the historically recorded change in temperature between 1860 and 1990. All current models suggest that global temperature will rise between now and 2050. Scientists admit that current models could underestimate or overestimate the amount of warming by a factor of two. (Data from National Academy of Sciences and National Center for Atmospheric Research)

to reach deeper layers. Recent measurements indicate that deep vertical mixing in the ocean occurs extremely slowly—taking up to hundreds of years—in most places. The oceans also help moderate tropospheric temperature by removing at least one-third of the excess carbon dioxide (a major greenhouse gas) we pump into the atmosphere, but we don't know if they can absorb more. If the oceans warm up enough, more carbon dioxide will bubble out of solution than dissolves (just as in a glass of ginger ale left out in the sun), amplifying and accelerating global warming. Another possibility is that warmer air will evaporate more water from the oceans and create more clouds. Depending on their type (thick or thin) and altitude, more clouds could contribute to either warming or cooling. We don't know which type might predominate and how this factor would vary in different parts of the world.

- *Changes in polar ice.* The Greenland and Antarctic ice sheets act like enormous mirrors reflecting sunlight back into space. If warmer temperatures melted some of this ice and exposed darker ground or ocean that would absorb more sunlight, warming would be accelerated. Then more ice would melt, amplifying the rise in atmospheric temperature even more. On the other hand, the early stages of global warming might actually increase the amount of the earth's water stored as ice. Warmer air would carry more water vapor, which could drop more snow on some glaciers, especially the gigantic Antarctic ice sheet. If snow accumulated faster than ice was lost, the ice sheet would grow, reflect more sunlight, and help cool the atmosphere—perhaps ushering in a new ice age within a thousand years.
- *Air pollution.* Projected global warming might be partially offset by particles and droplets of various air pollutants (released by volcanic eruptions and human activities) because they reflect back some of the incoming sunlight. However, things aren't that simple. Pollutants in the lower troposphere can either warm or cool the air and surface below them, depending on the reflectivity of the underlying surface. These contradictory and patchy effects, plus improved air pollution control, make it unlikely that air pollutants will counteract any warming very much in the next half century. Even if they did, levels of these pollutants, which already kill hundreds of thousands of people a year and damage vegetation (including food crops), need to be reduced.
- *Effects of increased CO_2 on photosynthesis.* Some studies suggest that more carbon dioxide in the atmosphere is likely to increase the rate of photosynthesis, with the increased growth of plants and other producers removing more carbon dioxide from the atmosphere and slowing global

Q: How much of the commercial energy used in the United States comes from nonrenewable resources?

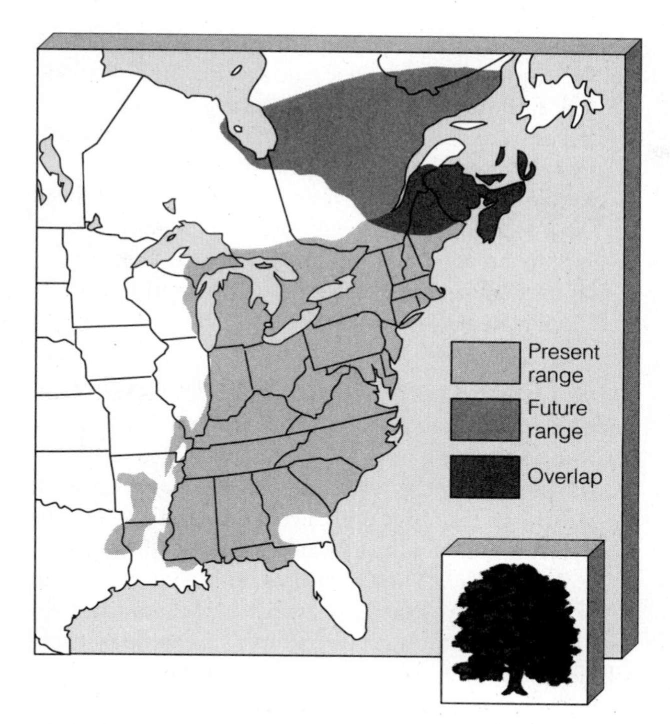

Figure 6-6 According to one projection, if carbon dioxide emissions doubled between 1990 and 2050, beech trees (now common throughout the eastern United States) would be able to survive only in a greatly reduced range in northern Maine and southeastern Canada. (Data from Margaret B. Davis and Catherine Zabinski, University of Minnesota)

warming (Figure 6-3). Other studies suggest that this effect varies with different types of plants. Also, much of the increased plant growth could be offset by plant-eating insects that breed more rapidly and year-round in warmer temperatures.

- *Methane release.* Some scientists speculate that in a warmer world huge amounts of methane tied up in arctic tundra soils and in muds on the bottom of the Arctic Ocean might be released if the blanket of permanently frozen ice (permafrost) in tundra soils melts and the oceans warm considerably. Because methane is a potent greenhouse gas, this release could greatly amplify global warming. On the other hand, some scientists believe that bacteria in tundra soils would rapidly oxidize the escaping methane to carbon dioxide, a less potent greenhouse gas.

Because of these and numerous other uncertainties in global climate models, projections made so far might be off by a factor of two in either direction. In other words, global warming during the next 50–100 years could be half the projected temperature increase in Figure 6-5 (the best-case scenario) or double it (the worst-case scenario).

SOME POSSIBLE EFFECTS OF A WARMER WORLD

So what's the big deal? Why should we worry about a possible rise of only a few degrees in the mean surface temperature of the earth? We often have that much change between June and July, or between yesterday and today. The key point is that we are not talking about normal swings in *local weather*, but about a possible *global* change in *average climate* from a thickening blanket of greenhouse gases in the troposphere.

A warmer troposphere would have different consequences for different peoples and species. Some places would get drier, some wetter. Some would get hotter, others cooler. Here are some possible effects of a warmer global climate:

- *Changes in food production.* Food productivity could increase in some areas and drop in others. Archeological evidence and computer models indicate that climate belts and thus tolerance ranges of plant species (including crops) would shift northward by 100–150 kilometers (60–90 miles) or 150 meters (500 feet) vertically up mountainsides for each 1°C (1.8°F) rise in global temperature. Computer models have projected drops in global yields of key food crops ranging from 30% to 70%. With current knowledge, we can't predict where changes in crop-growing capacity might occur or how long such changes might last.

- *Reductions in water supplies.* Lakes, streams, and aquifers in some areas that have watered ecosystems, croplands, and cities for centuries could shrink or dry up altogether, forcing entire populations to migrate to areas with adequate water supplies—if they could. We can't say with much certainty where this might happen.

- *Changes in forests.* Forests in temperate and subarctic regions would die out unless they could move toward the poles or to higher altitudes. However, tree species can move only through the slow growth of new trees along forest edges—typically about 0.9 kilometer (0.5 mile) per year or 9 kilometers (5 miles) per decade. If climate belts moved faster than this or if migration were blocked by cities, roads, or other barriers built by people, entire forests of oak, beech (Figure 6-6), and other deciduous (leaf-shedding) trees could die and release CO_2 into the atmosphere as they decompose. According to Oregon State University scientists, projected drying from global warming could cause massive fires in up to 90% of North American forests, destroying wildlife habitats and injecting into the atmosphere huge amounts of carbon dioxide, which could amplify global warming.

A: 91% (84% from fossil fuels and 7% from nuclear power)

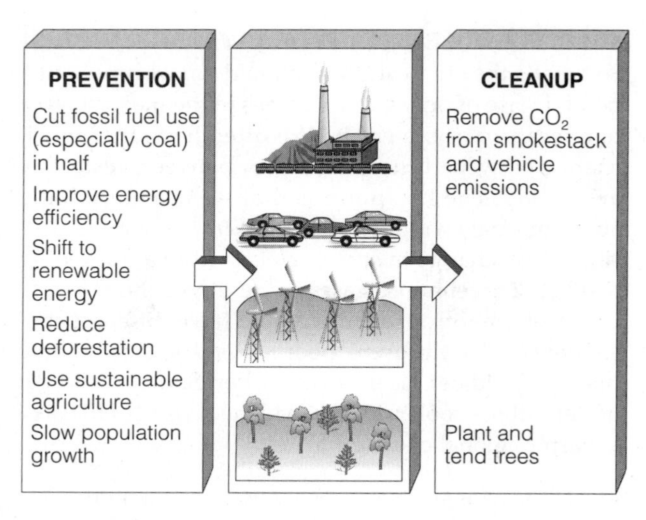

Figure 6-7 Methods for slowing possible global warming.

- *Reductions in biodiversity.* Large-scale forest diebacks would also cause mass extinction of species that couldn't migrate to new areas. And fish would die as temperatures soared in streams and lakes and as lowered water levels concentrated pesticides. Any shifts in regional climate would threaten many parks, wildlife reserves, wilderness areas, wetlands, and coral reefs, wiping out many current efforts to stem the loss of biodiversity (Chapter 4).
- *Rising sea level.* Water expands slightly when heated. This explains why global sea levels would rise if the oceans warmed, just as the fluid in a thermometer rises when heated. If warming at the poles caused ice sheets and glaciers to melt even partially, global sea level would rise even more. About one-third of the world's population and more than a third of the world's economic infrastructure are concentrated in coastal regions. Thus, even a moderate rise in sea level would flood low-lying areas (many of which contain major cities), as well as lowlands and deltas where crops are grown. It would also destroy wetlands and coral reefs and accelerate coastal erosion. One comedian jokes of plans to buy land in Kansas because it will probably become valuable beach-front property; another boasts she isn't worried because she lives in a houseboat—the "Noah strategy."
- *Weather extremes.* In a warmer world, prolonged heat waves and droughts would become the norm in many areas. And as the upper layers of seawater warmed, hurricanes and typhoons would occur more frequently and blow more fiercely.
- *Threats to in human health.* A warmer world would disrupt supplies of food and fresh water, displacing millions of people and altering disease patterns in unpredictable ways. The spread of tropical climates from the equator would bring malaria, encephalitis, yellow fever, dengue fever, and other insect-borne diseases to formerly temperate areas. A rise in sea level could spread infectious diseases by flooding sewage and sanitation systems in coastal cities.

SOLUTIONS: SLOWING GLOBAL WARMING

Some analysts believe we should wait until we know more before taking any serious action to deal with the possibility of global warming. However, even with better understanding, which could take decades, our knowledge will be limited because climate is so incredibly complex. Thus scientists will never be able to offer the certainty that some decision makers want before making such tough decisions as phasing out fossil fuels and replacing deforestation with reforestation.

Others urge us to adopt the *precautionary principle*, the idea that when dealing with risky, unpredictable, and often irreversible environmental problems it is often wise to take action before there is enough scientific knowledge to justify it.

Figure 6-7 presents a variety of ways to slow possible global warming; currently, none of these solutions is being vigorously pursued. The quickest, cheapest, and most effective way to reduce emissions of CO_2 and other air pollutants over the next two to three decades is to use energy more efficiently (Solutions, p. 115, and Section 5-2). According to the National Academy of Sciences, this strategy alone could lower U.S. greenhouse gas emissions by 10–40% at no net cost to the economy. Shifting from fossil fuels to renewable energy resources that do not emit carbon dioxide (Sections 5-3 and 5-4) could cut projected U.S. carbon dioxide emissions by 8–15% by 2000 and could virtually eliminate them by 2025.

Natural gas (Section 5-7) could be used to help make the transition to an age of energy efficiency and renewable energy. When burned, natural gas emits only half as much carbon dioxide per unit of energy as coal (Figure 5-15) and emits far smaller amounts of most other air pollutants as well. Halting deforestation (Sections 4-2 and 4-3) and switching to Earth-sustaining agriculture (Section 3-7) would reduce carbon dioxide emissions and help preserve biodiversity. Slowing population growth (Section 2-4) is also important. If we cut per capita greenhouse-gas emissions in half but world population doubles, we're back where we started.

Some analysts have suggested that MDCs and LDCs enter into win-win pacts to reduce the threat of

Q: How much of the commercial energy used in the United States is wasted?

global warming. In this "let's make a deal" strategy, LDCs would agree to stop deforestation, protect biodiversity, slow population growth, enact fairer land distribution policies, and phase out coal burning. In return, MDCs would forgive much of LDCs' foreign debt and help fund the transfer to LDCs of modern energy efficiency, solar energy, pollution control, pollution prevention, sustainable agriculture, and reforestation technologies. MDCs would also agree to make substantial cuts in their use of fossil fuels, abandon use of ozone-depleting chemicals, greatly improve energy efficiency, stop deforestation, shift to sustainable agriculture, and slow their population growth.

Removing significant amounts of carbon dioxide from exhaust gases is not currently feasible. Available methods can remove only about 30% of the carbon dioxide and would at least double the cost of electricity. Planting and tending trees is vital for restoring deforested and degraded land and for reducing soil erosion, but it is only a stopgap measure for slowing carbon dioxide emissions. To absorb the carbon dioxide we put into the atmosphere, each person in the world would need to plant and tend an average of 1,000 trees every year. Also, if much of the resulting newly grown forests are cleared and burned by us or by massive forest fires caused by global warming—or if much of the new forest died because of drought—most of the carbon dioxide removed would be released, accelerating global warming.

Some scientists have suggested various "technofixes" for dealing with possible global warming, including **(1)** fertilizing the oceans with iron to stimulate the growth of marine algae, which could remove more carbon dioxide through photosynthesis; **(2)** covering the oceans with white Styrofoam chips to help reflect more energy away from the earth's surface; **(3)** unfurling gigantic foil-faced sun shields in space to reduce solar input; and **(4)** injecting sunlight-reflecting sulfate particulates into the stratosphere to cool the earth's surface. All of these schemes are quite expensive and would have unknown—possibly harmful—effects on the earth's ecosystems and climate.

Energy Efficiency to the Rescue

SOLUTIONS

According to energy expert Amory Lovins, *the remedies for slowing possible global warming are things we should be doing already, even if there were no threat of global warming.* He also argues that getting countries to sign treaties and to agree to cut back and reallocate their use of fossil fuels in time to reduce serious environmental effects is difficult, if not almost impossible, and very costly. Climate models suggest that carbon dioxide emissions must be reduced by 80% to slow projected global warming to a safe rate; so far countries can't even agree to a 20% reduction.

According to Lovins, improving energy efficiency (Section 5-2) would be the fastest, cheapest, and surest way to slash emissions of carbon dioxide and most other air pollutants within two decades using existing technology. This approach would also save the world up to $1 trillion a year in reduced energy costs—as much as the annual global military budget.

Moreover, using energy more efficiently would reduce all forms of pollution, help protect biodiversity, and forestall arguments among governments about how carbon dioxide reductions should be divided up and enforced. This approach would also make the world's supplies of fossil fuel last longer, reduce international tensions over who gets the dwindling oil supplies, and give us more time to phase in alternatives to fossil fuels.

Lovins and other scientists argue that greatly improving worldwide energy efficiency *now* is a money-saving, life-saving, Earth-saving offer that we should not refuse, even if climate change was not a possibility. What do you think?

SOLUTIONS: PREPARING FOR GLOBAL WARMING Some analysts suggest that we begin preparing for the long-term effects of possible global warming. Their suggestions include the following courses of action:

- *Breed food plants that need less water or can thrive in salty water.*
- *Build dikes to protect coastal areas from flooding,* as the Dutch have done for centuries.
- *Move storage tanks of hazardous materials away from coastal areas.*
- *Ban new or rebuilt construction on low-lying coastal areas.*
- *Stockpile 1–5 years' worth of key foods throughout the world as short-term insurance against disruptions in food production.*
- *Expand existing wilderness areas, parks, and wildlife refuges* northward in the Northern Hemisphere and southward in the Southern Hemisphere, and create new wildlife reserves in these areas.
- *Connect wildlife reserves with corridors that would allow mobile species to move with climate change.*
- *Waste less water* (Section 7-3).

A: 84% (43% of this energy is unnecessarily wasted)

What *You* Can Do to Reduce Global Warming

- *Reduce your use of fossil fuels.* Driving a car that gets at least 15 kilometers per liter (35 miles per gallon), joining a car pool and using mass transit, and walking or bicycling as much as possible will reduce emissions of carbon dioxide and other air pollutants, will save energy and money, and can improve your health.
- *Use energy-efficient light bulbs, refrigerators, and other appliances* (Section 5-2).
- *Use solar energy to heat household space or water as much as possible* (Section 5-3).
- *Cool your house by using shade trees and available breezes and by making it energy efficient.*
- *Plant and care for trees to help absorb carbon dioxide.*

Each of us can play a role in reducing the threat of global warming and in the process can reduce resource waste, pollution, and loss of biodiversity (Individuals Matter, above).

Ozone Depletion: Serious Threat or a Hoax?

THE THREAT: LETTING IN DEADLY RAYS Thanks to the evolution of photosynthetic, oxygen-producing bacteria, the earth has had a stratospheric global sunscreen—the ozone layer—for the past 450 million years. However, considerable evidence indicates that we are thinning this screen with our recent use of chlorine- and bromine-containing compounds.

A few scientists dismiss the threat of ozone depletion, but the overwhelming consensus of researchers in this field is that ozone depletion by chemicals we have released into the atmosphere is a real threat to various forms of life on Earth—including humans.

CHLOROFLUOROCARBONS: FROM DREAM CHEMICALS TO NIGHTMARE CHEMICALS How did we get into this situation? It started when Thomas Midgley, Jr., a General Motors chemist, discovered the first chlorofluorocarbon (CFC) in 1930, and chemists then made similar compounds to create a family of highly useful CFCs.

These amazingly useful, chemically stable, odorless, nonflammable, nontoxic, and noncorrosive compounds seemed to be dream chemicals. Cheap to make, they became popular as coolants in air conditioners and refrigerators, propellants in aerosol spray cans, cleaners for electronic parts such as computer chips, sterilants for hospital instruments, fumigants for granaries and ship cargo holds, and building blocks for the bubbles in Styrofoam (used for insulation and packaging).

But it was too good to be true. In 1974 chemists Sherwood Rowland and Mario Molina made calculations indicating that CFCs were creating a global time bomb by lowering the average concentration of ozone in the stratosphere. They shocked both the scientific community and the $28-billion-per-year industry that makes these chemicals by calling for an immediate ban of CFCs in spray cans.

Here's what Rowland and Molina found: Spray cans, discarded or leaky refrigeration and air-conditioning equipment, and the production and burning of plastic foam products release CFCs into the atmosphere. These molecules are too unreactive to be removed, and (mostly through convection and random drift) they rise slowly into the stratosphere, taking 10–20 years to make the journey. There, under the influence of high-energy ultraviolet (UV) radiation, they break down and release chlorine atoms, which speed up the breakdown of ozone (O_3) into O_2 and O and cause ozone to be destroyed faster than it is formed. Each CFC molecule can last in the stratosphere for 65–110 years. During that time, each chlorine atom in these molecules—like a gaseous Pac-Man—can convert as many as 100,000 molecules of O_3 to O_2. If current models and measurements are correct, these one-time dream molecules have turned into a nightmare of global ozone terminators. Although Rowland and Molina warned us of this problem in 1974, it took 15 years of interaction between the scientific and political communities before countries agreed to begin slowly phasing out CFCs.*

CFCs are not the only ozone-eaters. A few other chemicals can release highly reactive chlorine and bromine atoms if they end up in the stratosphere and are exposed to intense UV radiation, including **(1)** bromine-containing compounds called *halons* and *HBFCs* (both used in fire extinguishers) and *methyl bromide* (a widely used pesticide); and **(2)** *carbon tetrachlo-*

* For a fascinating account of how corporate stalling, politics, economics, and science can interact, see Sharon Roan's *Ozone Crisis: The 15-Year Evolution of a Sudden Global Emergency* (New York: Wiley, 1989).

Q: What is the most inefficient and costly way to produce electricity for heating an interior space or water?

ride (a cheap, highly toxic solvent) and *methyl chloroform*, or 1,1,1-trichloroethane (used as a cleaning solvent for clothes and metals and as a propellant in more than 160 consumer products, such as correction fluid, dry-cleaning sprays, spray adhesives, and other aerosols). Substitutes are available for virtually all the uses of these chemicals.

HOLES IN THE OZONE LAYER AND OTHER SURPRISES Each year the news about ozone loss seems to worsen, and sometimes it takes scientists by surprise. The first surprise came in 1985, when researchers analyzing satellite data discovered that 50% of the ozone in the upper stratosphere over the Antarctic (98% in some localized areas) was being destroyed during the Antarctic spring and early summer (September–December), when sunlight returned after the dark Antarctic winter. This pronounced loss of ozone had not been predicted by computer models of the stratosphere. Since then, this seasonal Antarctic *ozone hole*, or *ozone thinning*, has expanded in most years. In both 1992 and 1993 it covered an area roughly three times the size of the continental United States.

Measurements have indicated that CFCs are the primary culprits. After weeks of ozone depletion once sunlight returns to the Antarctic, huge clumps of ozone-depleted air flow northward and linger for a few more weeks over parts of Australia and New Zealand, and over the southern tips of South America and Africa; this raises UV radiation levels in these areas by as much as 20%.

In 1988 scientists discovered that similar but much less severe ozone thinning occurs over the North Pole during the Arctic spring and early summer (February–June), with a seasonal ozone loss of 10–25% (compared to 50% or more over much of the Antarctic region). When this mass of air above the Arctic breaks up each spring, masses of ozone-depleted air flow southward to linger over parts of Europe, North America, and Asia. Mostly because these air masses flow alternately over land and water, seasonal ozone loss over the North Pole is lower than that over the South Pole.

The situation could get much worse. In 1992 atmospheric scientists warned that if rising levels of greenhouse gases change the climate as projected over the next 50 years, the stratosphere over the Arctic could be altered such that it will experience severe ozone thinning like that now found over Antarctica. This in turn would sharply decrease ozone levels over parts of the northern hemisphere, including the United States.

IS OZONE DEPLETION A HOAX? Political talk-show host Rush Limbaugh, zoologist and former head of the Atomic Energy Commission Dixy Lee Ray, and several articles in the popular press have claimed that ozone depletion by CFCs is an overblown hoax. The evidence for these claims comes mostly from articles and books written by S. Fred Singer (a Ph.D. physicist and climate scientist), Rogelio Maduro (who has a bachelor of science degree in geology and is an associate editor of a science and technology magazine published by supporters of Lyndon LaRouche—considered an extremist politician by most people), and Ralf Schauerhammer (a German writer).

Maduro and Schauerhammer claim that the evidence indicates that volcanoes, seawater, and biomass burning have been releasing far more chlorine into the atmosphere than do the CFCs we have emitted. They argue that this has been going on for billions of years and the ozone layer is still here.

Scientists directly involved in ozone-layer research dispute these claims, describing them as being based on selective use of out-of-date research and on unwarranted extrapolation of questionable data. They point out that most chlorine from natural sources—mostly sodium chloride (NaCl) and hydrogen chloride (HCl) from the evaporation of sea spray—never makes it to the stratosphere because these compounds (unlike CFCs) are soluble in water and get washed out of the lower atmosphere by rain. If sodium chloride from sea spray were making it to the stratosphere, there should be evidence of sodium in the lower stratosphere; measurements show that it is not there. Measurements also indicate that no more than 20% of the chlorine from biomass burning (in the form of methylchloride) is making it to the stratosphere, which is about five times less than the contribution from CFCs.

Researchers also dispute the hypothesis that large quantities of HCl are injected into the stratosphere from volcanic eruptions. Most of this water-soluble HCl is injected into the troposphere and is washed out by rain before it reaches the stratosphere. Measurements show that HCl in the stratosphere increased by less than 10% after the eruption of the Mexican volcano El Chichón in 1982, and it increased even less from the eruption of Mt. Pinatubo in the Philippines in 1991. Singer, whose skepticism about some aspects of ozone depletion models has been cited to bolster the case of those calling the whole thing a hoax, agrees with this scientific consensus and in 1993 stated that "CFCs make the major contribution to stratospheric chlorine."

Critics of the ozone-depletion idea also point out that the expected increase in UV radiation from ozone loss in the stratosphere has not as yet been detected in urban areas in the United States and most other MDCs. However, it is suspected that the reason for

A: Nuclear power (only 14% efficient)

this observation is that air pollution over urban areas is providing some protection against increased UV radiation at ground level—another example of connections. If such air pollution is decreased (Section 6-5), we may experience a greater threat from increased UV radiation; however, if we don't decrease air pollution, we will continue to suffer from its harmful effects (Section 6-4).

POSSIBLE CONSEQUENCES OF OZONE LOSS: LIFE IN THE ULTRAVIOLET ZONE Why should we care about ozone loss? With less ozone in the stratosphere, more biologically harmful ultraviolet radiation will reach the earth's surface. Evidence indicates that increased UV radiation will give us worse sunburns, earlier wrinkles, more cataracts (a clouding of the lens that reduces vision and can cause blindness if not corrected), and more skin cancers. Cases of skin cancer and cataracts are soaring in Australia, New Zealand, South Africa, Argentina, and Chile, where the ozone layer is very thin for several months after the masses of ozone-depleted air over the South Pole drift northward. Occurrences of skin cancer and cataracts are also increasing in the United States.

Other effects from increased UV exposure are **(1)** suppression of the immune system, which would reduce our defenses (regardless of skin pigmentation) against a variety of infectious diseases; **(2)** an increase in eye-burning, highly damaging ozone and acid deposition in the troposphere (Section 6-4); **(3)** lower yields (about a 1% decline for each 3% drop in stratospheric ozone) of crops such as corn, rice, soybeans, cotton, beans, peas, sorghum, and wheat; and **(4)** a loss of perhaps $2 billion per year from degradation of paints, plastics, and other polymers in the United States alone.

INDIVIDUALS MATTER

Ray Turner and His Refrigerator

Ray Turner, an aerospace manager at Hughes Aircraft in California, made an important low-tech, ozone-saving discovery by using his head—and his refrigerator. His concern for the environment led him to look for a cheap and simple substitute for the CFCs used as cleaning agents for removing films of oxidation formed during the manufacture of most electronic circuit boards at his plant and elsewhere.

He started his search by looking in his refrigerator for a better circuit-board cleaner. He decided to put drops of various substances on a corroded penny to see whether any of them would remove the film of oxidation. Then he used his soldering gun to see if solder would stick to the cleaned surface of the penny, indicating that the film had been cleaned off.

First, he tried vinegar. No luck. Then he tried some ground-up lemon peel, also a failure. Next he tried a drop of lemon juice and watched as the solder took hold. The rest is history.

In the months that followed Turner and a Hughes team perfected his discovery. Since it was introduced, the new cleaning technique has reduced circuit-board defects by about 75% at Hughes. And Turner got a hefty bonus. Maybe you can find a solution to an environmental problem in *your* refrigerator.

In a worst-case scenario, most people would have to avoid the sun altogether (see cartoon). Even cattle could graze only at dusk, and farmers and other outdoor workers might need to limit their exposure to the sun to minutes. One comic has said that he wasn't worried because the big hole in the ozone layer will allow the smoke from burning off the rain forests to go through.

SOLUTIONS: PROTECTING THE OZONE LAYER The scientific consensus is that we should stop producing ozone-depleting chemicals now (global abstinence). After saying no, we will still have to wait about 100 years for the ozone layer to return to 1985 levels, and another 100–200 years for full recovery.

Substitutes are already available for most uses of CFCs, and others are being developed (Individuals

Q: What are the two most energy-efficient ways to heat interior space?

Matter, left). CFCs in existing air conditioners, refrigerators, and other products must be recovered and in some cases reused until the substitutes are phased in. This will be expensive, but if ozone depletion should become more severe because of our activities, the ecological, health, and financial costs will be much higher.

Substitutes such as hydrofluorocarbons (HFCs) and hydrochlorofluorocarbons (HCFCs) may help ease the replacement of CFCs for essential uses, but models indicate that these chemicals also can deplete ozone (although at a slower rate) and should be banned no later than 2005.

CAN TECHNOFIXES SAVE US? What about a quick fix from technology so we can keep on using CFCs? One suggestion is to collect some of the ozone-laden air at ground level over Los Angeles and other cities and ship it up to the stratosphere. Even if we knew how to do this at an affordable cost, the Los Angeles air would dilute the stratospheric ozone concentration, rather than increasing it.

Two atmospheric scientists have speculated that we might inject large quantities of ethane and propane into the stratosphere, where they might react with CFCs to remove the offending chlorine atoms. But the scientists proposing this possibility warn that the plan could backfire, accelerate ozone depletion, and have unpredictable effects on climate.

Others have suggested using tens of thousands of lasers to blast CFCs out of the atmosphere before they can reach the stratosphere. However, the energy requirements would be enormous and expensive, and decades of research would be needed to perfect the types of lasers needed. And no one can predict the possible effects on climate, birds, or planes.

SOME HOPEFUL PROGRESS In 1987, 24 nations meeting in Montreal developed a treaty—commonly known as the *Montreal Protocol*—to cut emissions of CFCs (but not other ozone depleters) into the atmosphere by about 35% between 1989 and 2000. After hearing more bad news about ozone depletion, representatives of more than half the world's nations met in Copenhagen in 1992 and agreed to **(1)** phase out production (except for essential uses) of CFCs, carbon tetrachloride, halons, HBFCs (halon substitutes), and chloroform by January 1, 1996; **(2)** freeze consumption of HCFCs at 1991 levels by 1996 and eliminate them by 2030; and **(3)** freeze methyl bromide production at 1991 levels by 1995.

The agreements reached so far are important examples of global cooperation. Some scientists claim that such action is premature because models of ozone depletion and its effects are not reliable enough. But many other scientists believe that ozone loss is a serious threat and that the agreements do not go far enough fast enough. Each of us can play a role in reducing the threat of ozone depletion (Individuals Matter, above).

What *You* Can Do to Help Protect the Ozone Layer

If you believe, as do most scientists, that there is a serious threat to the ozone layer, then follow these suggestions:

- *Don't buy products containing CFCs, carbon tetrachloride, or methyl chloroform* (1,1,1-trichloroethane on most ingredient labels).
- *Don't buy CFC-containing polystyrene foam insulation.* Types of insulation that don't contain CFCs are extended polystyrene (commonly called EPS or beadboard), fiberglass, rock wool, cellulose, and perlite.
- *Don't buy halon fire extinguishers for home use.* Instead, buy those that use dry chemicals (carbon dioxide extinguishers release this greenhouse gas into the troposphere). If you already have a halon extinguisher, store it until a halon-reclaiming program is developed.
- *Stop using aerosol spray products*, except in some necessary medical sprays. Even those that don't use CFCs and HCFCs (such as Dymel) emit hydrocarbons or other propellant chemicals into the air. Use roll-on and hand-pump products instead.
- *Pressure legislators to ban all CFCs, halons, methyl bromide, carbon tetrachloride, and methyl chloroform by 1996 (with no loopholes) and HCFCs by 2005 instead of by 2030.*
- *Pressure legislators not to exempt military and space programs from any phaseout of ozone-depleting chemicals.*
- *If you junk a car, a refrigerator, a freezer, or an air conditioner, make sure the coolant is removed and kept safely for reuse or destruction.*
- *Have car and home air conditioners checked regularly for CFC leaks and repair them if necessary.*
- *If you buy a car with an air conditioner, look for one that doesn't use CFCs.* These should be available on some models in 1994 and on most models by 1995.

A: Passive solar and natural gas

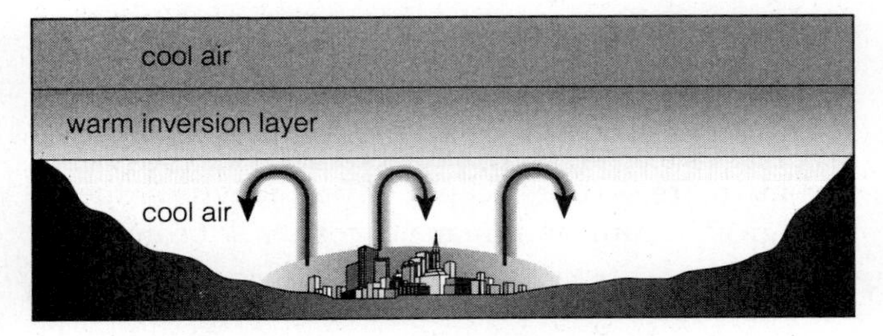

Figure 6-8 Thermal inversions trap pollutants in a layer of cool air that cannot rise to carry the pollutants away.

6-4 Air Pollution

OUTDOOR AIR POLLUTION As clean air in the troposphere moves across the earth's surface, it collects the products of both natural events (volcanic eruptions and dust storms) and human activities (emissions from cars and smokestacks). These potential pollutants, called **primary pollutants**, mix with the churning air in the troposphere. Some may react with one another or with the basic components of air to form new pollutants, called **secondary pollutants**. Long-lived pollutants travel far before they return to the earth as particles, droplets, or chemicals dissolved in precipitation.

Outdoor pollution in industrialized countries comes mostly from five groups of primary pollutants: carbon oxides (CO and CO_2), nitrogen oxides (mostly NO and NO_2, or NO_x), sulfur oxides (SO_2 and SO_3), volatile organic compounds (mostly hydrocarbons), and suspended particles and droplets (particulates), all produced primarily by combustion of fossil fuels (Section 5-7).

In MDCs, most pollutants enter the atmosphere from the burning of fossil fuels both in power plants and factories (*stationary sources*) and in motor vehicles (*mobile sources*). In car-clogged cities like Los Angeles, São Paulo, London, and Mexico City, motor vehicles are responsible for 80–88% of the air pollution. Two important types of outdoor air pollution are smog and acid deposition.

CONNECTIONS: SMOG *Photochemical smog* is a mixture of primary and secondary pollutants that forms when some of the primary pollutants interact under the influence of sunlight. The resulting mix of more than 100 chemicals is dominated by ozone, a highly reactive gas that harms most living organisms.

Virtually all modern cities have photochemical smog, but it is much more common in cities with sunny, warm, dry climates and lots of motor vehicles. Los Angeles, Denver, Salt Lake City, Sydney, Mexico City, and Buenos Aires all have serious photochemical smog problems. The hotter the day, the higher the levels of ozone and other components of photochemical smog.

Thirty years ago cities like London, Chicago, and Pittsburgh burned large amounts of coal and heavy oil (which contain sulfur impurities) in power plants and factories, as well as for space heating. During winter such cities suffered from *industrial smog*, consisting mostly of a mixture of sulfur dioxide, suspended droplets of sulfuric acid (formed from some of the sulfur dioxide), and a variety of suspended solid particles. Today, in most parts of the world coal and heavy oil are burned only in large boilers with reasonably good pollution control or with tall smokestacks, so industrial smog, sometimes called *gray-air smog*, is rarely a problem. However, in China, Ukraine, and some eastern-European countries large quantities of coal are still burned with inadequate pollution controls.

The frequency and severity of smog in an area depend on several things: the local climate and topography, the density of the population, the amount of industry, and the fuels used in industry, heating, and transportation. In areas with high average annual precipitation, rain and snow help cleanse the air of pollutants. Winds also help sweep pollutants away and bring in fresh air, but they may transfer some pollutants to distant areas.

Hills and mountains tend to reduce the flow of air in valleys below and allow pollutant levels to build up at ground level. Buildings in cities generally slow wind speed, thereby reducing dilution and removal of pollutants.

During the day the sun warms the air near the earth's surface. Normally this heated air expands and rises, carrying low-lying pollutants higher into the troposphere. Colder, denser air from surrounding high-pressure areas then sinks into the low-pressure area created when the hot air rises (Figure 6-8, left). This continual mixing of the air helps keep pollutants from reaching dangerous concentrations near the ground.

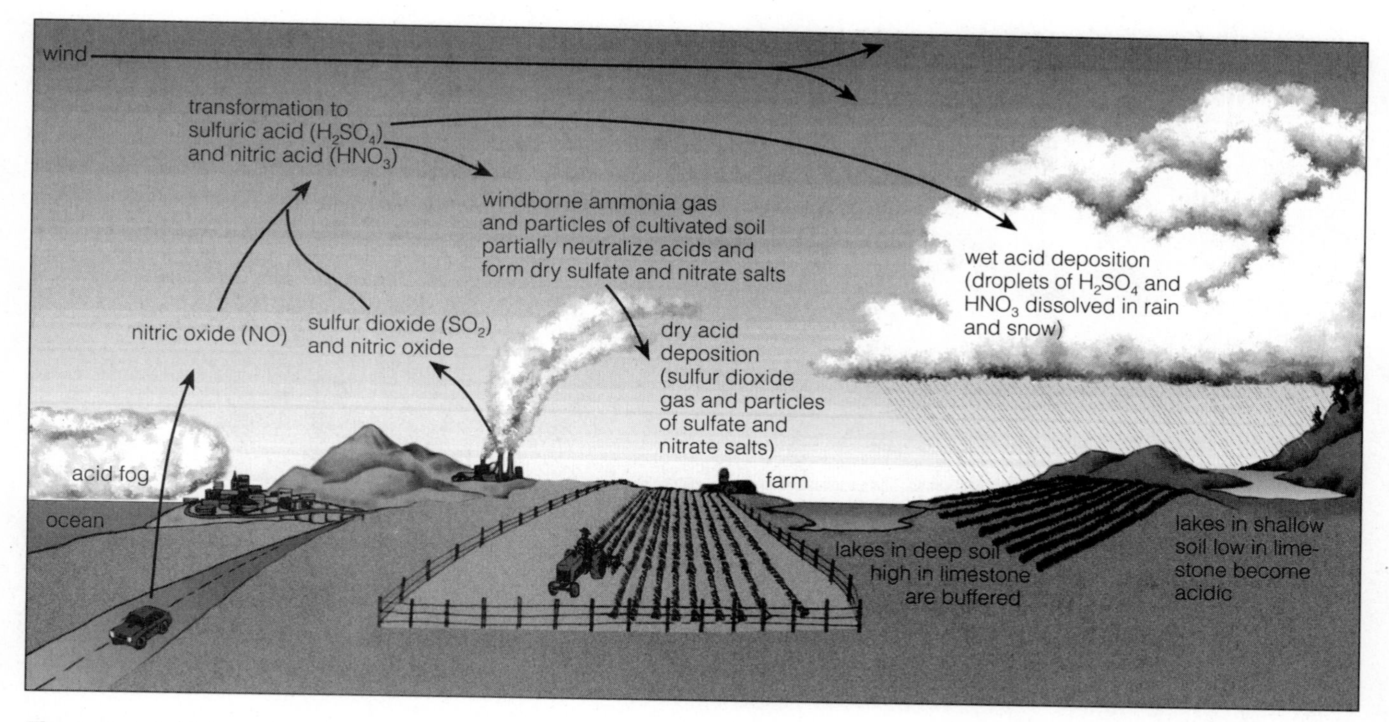

Figure 6-9 Acid deposition consists of acidified rain, snow, dust, or gas.

Sometimes, however, weather conditions trap a layer of dense, cool air beneath a layer of less dense, warm air in an urban basin or valley, a phenomenon called a *temperature inversion* or a *thermal inversion* (Figure 6-8, right). In effect, a lid of warm air covers the region and prevents ascending air currents (that would disperse pollutants) from developing. Such inversions usually last for only a few hours; but sometimes, when a high-pressure air mass stalls over an area, they last for several days. Then air pollutants at ground level build up to harmful and even lethal concentrations.

The first U.S. air pollution disaster occurred in 1948, when fog laden with sulfur dioxide and suspended particulate matter stagnated for five days over the town of Donora in the Monongahela Valley south of Pittsburgh, Pennsylvania. About 6,000 of the town's 14,000 inhabitants fell ill, and 20 of them died. This killer fog resulted from a combination of mountainous terrain surrounding the valley and stable weather conditions. These conditions trapped and concentrated deadly pollutants emitted by the community's steel mill, zinc smelter, and sulfuric acid plant.

A city with several million people and automobiles in an area with a sunny climate, light winds, mountains on three sides, and the ocean on the other has ideal conditions for photochemical smog worsened by frequent thermal inversions. This describes the Los Angeles basin, which has 14 million people, 10.6 million motor vehicles, thousands of factories, and thermal inversions at least half of the year. Despite having the world's toughest air pollution control program, Los Angeles is the air pollution capital of the United States. By 2010, the Los Angeles area is expected to have 21 million people and 13 million cars.

CONNECTIONS: ACID DEPOSITION To reduce local air pollution (and meet government standards without having to add expensive air pollution control devices), coal-burning power plants, ore smelters, and other industrial plants began using tall smokestacks to emit sulfur dioxide, suspended particles, and nitrogen oxides above the inversion layer. As this practice spread in the 1960s and 1970s, pollution in downwind areas began to rise. In addition to smokestack emissions, large quantities of nitrogen oxides are also released by motor vehicles.

As sulfur dioxide and nitrogen oxides are transported as much as 1,000 kilometers (600 miles) by prevailing winds, they form secondary pollutants such as nitric acid vapor, droplets of sulfuric acid, and particles of sulfate and nitrate salts (Figure 6-9). These chemicals descend to the earth's surface in two forms: *wet*, as acidic rain, snow, fog, and cloud vapor, and *dry*, as acidic particles. The resulting mixture is called **acid deposition**, commonly called *acid rain*. Because these acidic components remain in the atmosphere for only a few days, acid deposition occurs on a regional, rather than a global, basis.

A: Improve energy efficiency (especially of motor vehicles)

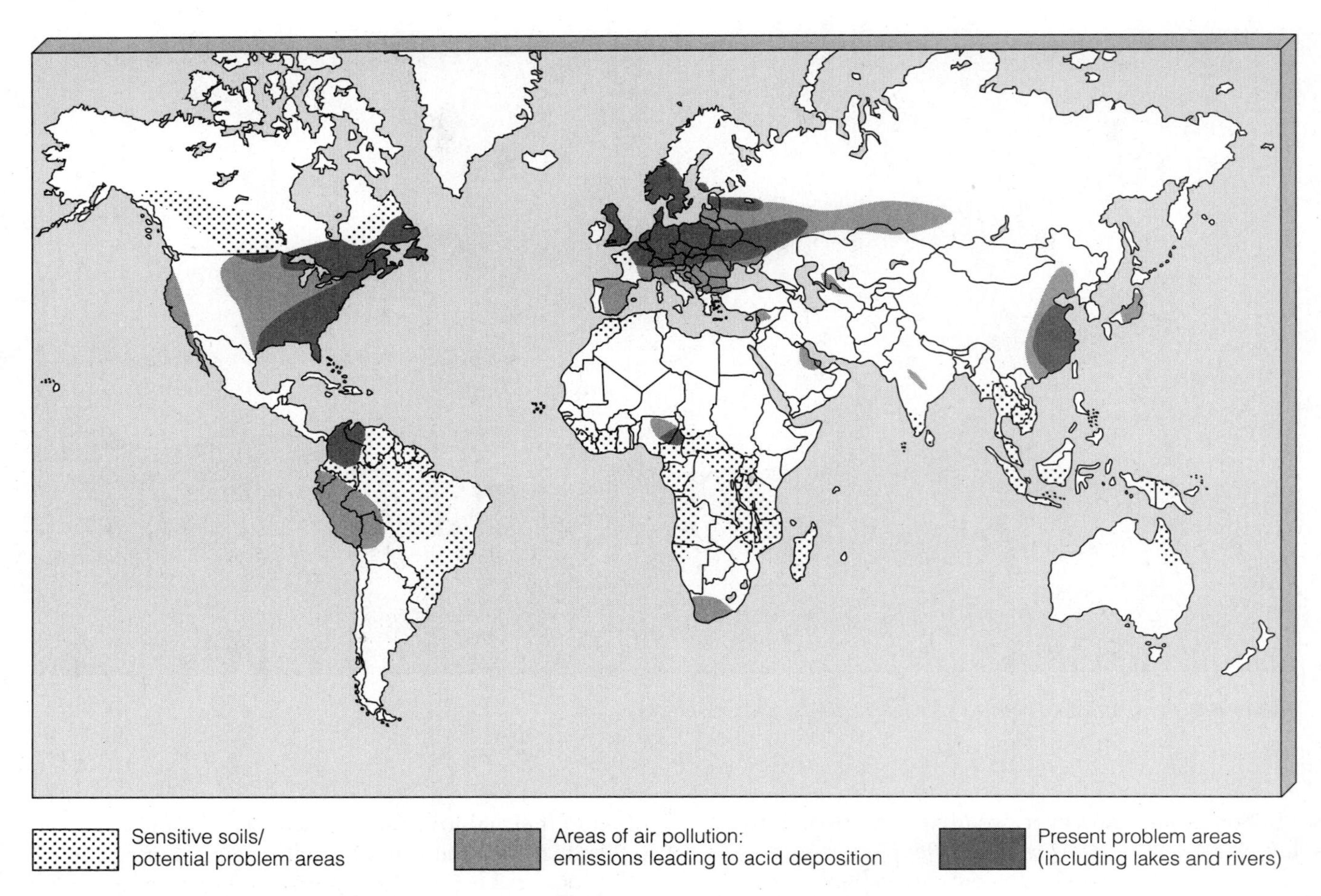

Figure 6-10 Regions where acid deposition is now a problem, and regions with potential acid deposition problems. (Data from World Resource Institute and Environmental Protection Agency)

Acid deposition has a number of harmful effects, including the following:

- It damages statues, buildings, metals, and car finishes.
- It can contaminate fish in some lakes with highly toxic methylmercury. Apparently, increased acidity of lakes converts inorganic mercury compounds in lake-bottom sediments into methylmercury, which is more soluble in the fatty tissue of animals.
- It can damage foliage and weaken trees, especially conifers such as red spruce at high elevations (which are bathed almost continuously in very acidic fog and clouds).
- It and other air pollutants can make trees more susceptible to stresses such as cold temperatures, diseases, insects, drought, and fungi that thrive under acidic conditions.
- It can release soluble aluminum ions from soil, which damage tree roots. When washed into lakes, the released aluminum ions also can kill many kinds of fish by increasing mucus formation, which clogs their gills.
- It leads to excessive levels of nitrogen in the soil, which can overstimulate plant growth and increase depletion of other soil nutrients.
- It contributes to human respiratory diseases such as bronchitis and asthma, which can cause premature death.

Acid deposition is a serious regional problem (Figure 6-10) in many areas downwind from coal-burning power plants, smelters, factories, and large urban areas. A large portion of the acid-producing chemicals generated in one country may be exported to others by prevailing winds. For example, more than three-fourths of the acid deposition in Norway, Switzerland, Austria, Sweden, the Netherlands, and Finland is blown to those countries from industrialized areas of western and eastern Europe.

Chemical detective work indicates that more than half the acid deposition in southeastern Canada and

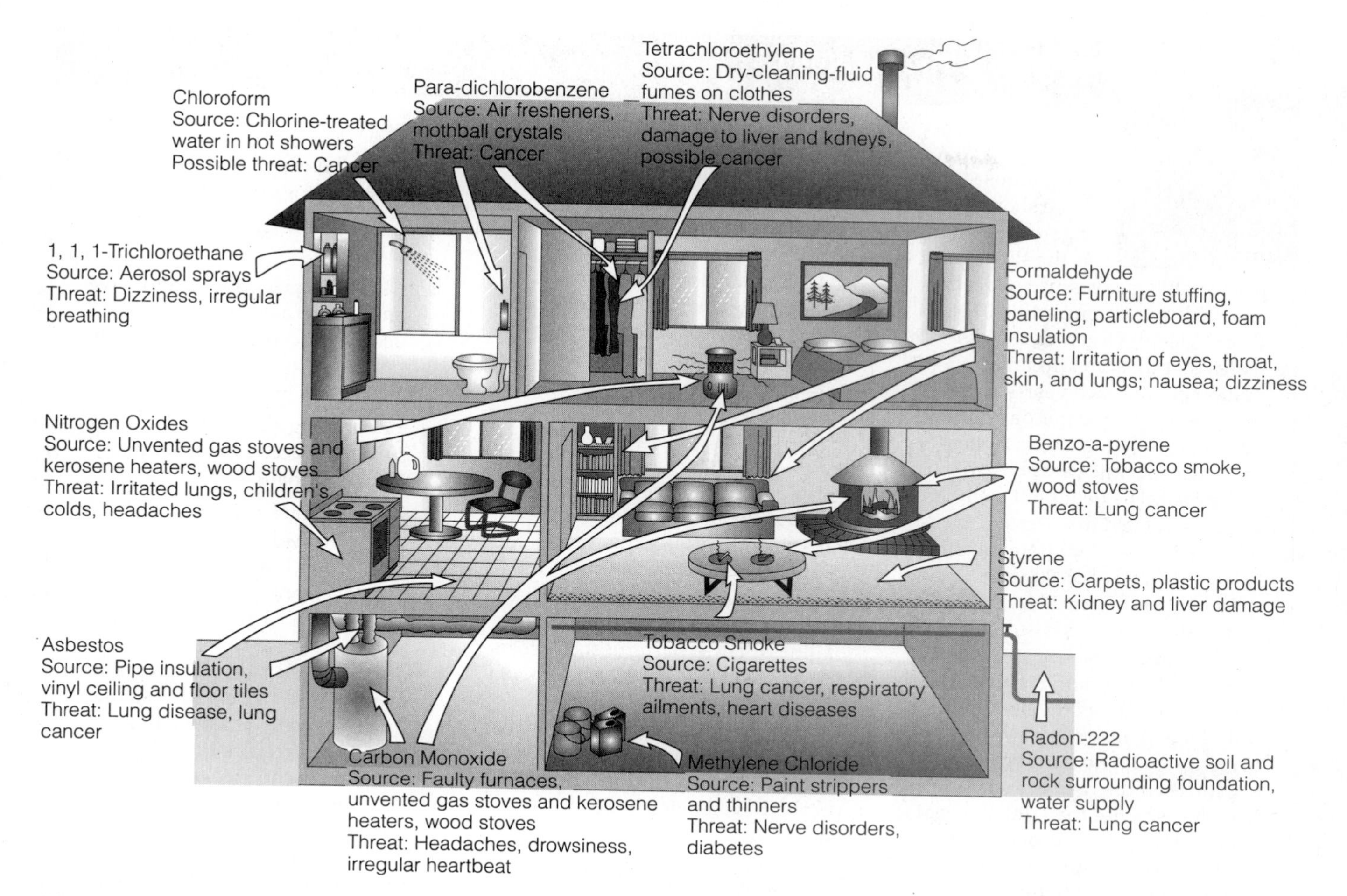

Figure 6-11 Some important indoor air pollutants. (Data from Environmental Protection Agency)

the eastern United States originates from coal- and oil-burning power plants and factories in seven states: Ohio, Indiana, Pennsylvania, Illinois, Missouri, West Virginia, and Tennessee. In areas near and downwind from large urban areas, emissions of nitrogen oxides (mostly from motor vehicles) leading to the formation of nitric acid may be the main culprit.

A large-scale, government-sponsored research study on acid deposition in the United States in the 1980s concluded that the problem was serious but not yet at a crisis stage. Representatives of coal companies and of industries that burn coal and oil claim that adding expensive air pollution control equipment or burning low-sulfur coal or oil costs more than the resulting health and environmental benefits are worth. However, a 1990 economic study indicated that the benefits of controlling acid deposition will be worth at least $5 billion (some say $10 billion) per year, about 50% greater than the costs of controlling acid deposition.

There is some good news. A 1993 study by the U.S. Geological Survey found that the concentration of sulfate ions—a key component of acid deposition—declined at 26 out of 33 of the U.S. rainwater collection sites between 1980 and 1991. And between 1970 and 1992, emissions of sulfur dioxide in the United States have dropped 30%. The 1990 amendments to the Clean Air Act require the nation to reduce its sulfur dioxide output to 60% of the 1980 level by 2000.

INDOOR AIR POLLUTION If you are reading this book indoors, you may be inhaling more air pollutants with each breath than if you had been outside (Figure 6-11). The health risks from exposure to such chemicals are magnified because people spend 70–98% of their time indoors. In 1990, the EPA placed indoor air pollution at the top of the list of 18 sources of cancer risk. At greatest risk are smokers, the young, the old, the sick, pregnant women, people with respiratory or heart problems, and factory workers. A 1993 study found that pollution levels inside cars can be up to 18 times higher than those outside the vehicles.

Pollutants found in buildings produce dizziness, headaches, coughing, sneezing, nausea, burning eyes, chronic fatigue, and flulike symptoms—the "sick

A: $1 trillion a year

Smoking: The Big Killer

Numerous studies have shown that cigarette smoking is the single most preventable major cause of death and suffering among adults. The World Health Organization estimates that tobacco kills at least 3 million people each year—an average of 8,200 people per day—from heart disease, lung cancer, other cancers, bronchitis, emphysema, and stroke. The death toll from smoking-related diseases in 2050 is projected to be 12 million annually—an average of 33,000 preventable deaths per day.

In 1992, smoking killed about 435,000 Americans—an average of 1,190 deaths per day (Figure 6-12). This death toll is equivalent to three fully-loaded jumbo jets crashing every day with no survivors. Smoking causes more deaths each year in the United States than do all illegal drugs, alcohol (the second most harmful drug), automobile accidents, suicide, and homicide combined. Studies show that passive smoke (inhaled by nonsmokers) is also a killer, causing up to 40,000 premature deaths in the United States a year.

Smoking is also highly addictive. A British government study showed that adolescents who smoke more than one cigarette have an 85% chance of becoming smokers. About 75% of smokers who quit start smoking again within six months, about the same relapse rate as for recovering alcoholics and heroin addicts.

Smoking costs the United States at least $65 billion per year (perhaps $95 billion) in expenses related to premature death, disability, medical treatment, increased insurance costs, and lost productivity because of illness. These harmful social costs amount to at least $3.00 per pack of cigarettes sold.

To reduce this hazard to human health, the American Medical Association and many health experts advocate the following actions:

- Banning all cigarette advertising in the United States.
- Forbidding the sale of cigarettes and other tobacco products to anyone under 21, with strict penalties for violators.
- Banning all cigarette vending machines.
- Classifying nicotine as addictive and dangerous and placing its use in tobacco or other products under the jurisdiction of the Food and Drug Administration.
- Eliminating all federal subsidies to U.S. tobacco farmers and tobacco companies.
- Taxing cigarettes at about $3.00 a pack (instead of the current 51¢, the second lowest rate in the world; cigarettes are taxed at $4 per pack in Norway and at more than $3 per pack in Denmark, Canada, and Great Britain). This action would discourage smoking (especially among adolescents) and require that people who choose to smoke pay for the resulting harmful costs now borne by society as a whole—a user-pays approach. Such a program would raise about $50 billion per year in revenue that could be used to help finance a better health-care system.
- Forbidding U.S. officials to pressure other governments to import U.S. tobacco or tobacco products. Since 1985 the federal government has threatened trade sanctions against countries that place tariffs and other restrictions on U.S. tobacco products. Thus the U.S. government is coercing other governments into allowing imports of a hazardous (but legal) drug from America while trying to halt the flow of illegal drugs into the United States.

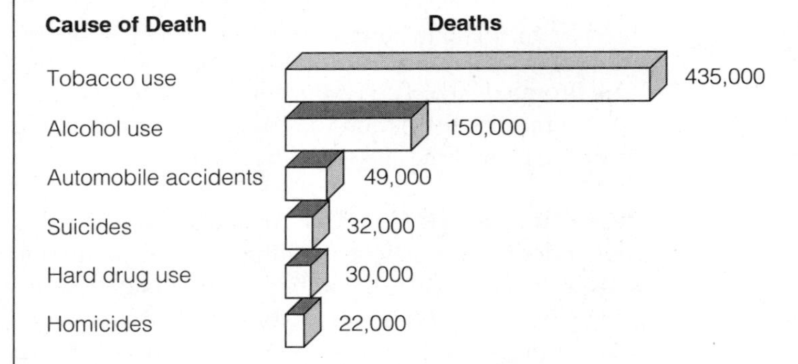

Figure 6-12 Deaths in the United States from tobacco use and other causes in 1992. Smoking is by far the nation's leading cause of preventable death, causing more premature deaths each year than all the other categories in this figure combined. (Data from National Center for Health Statistics)

Q: What is the largest untapped energy source in the United States?

building syndrome." According to the EPA, at least one-fifth of all U.S. buildings are considered "sick," at a cost to the nation of an estimated $60 billion per year in absenteeism and reduced productivity. Some indoor pollutants cause disease and premature death. According to the EPA and public health officials, cigarette smoke (Spotlight, left), radioactive radon-222 gas (which is released by certain types of rock and can seep into houses), tiny fibers of asbestos released indoors from building materials used between 1900 and 1986, and formaldehyde (used in plywood, particleboard, paneling, fiberboard, and carpeting and wallpaper adhesives) are the four most dangerous indoor air pollutants.

Severe indoor air pollution, especially from particulate matter, occurs inside the dwellings of many poor rural people in LDCs. The burning of wood, dung, and crop residues in open fires or in unvented or poorly vented stoves for cooking and heating (in temperate and cold areas) exposes the people, especially women and young children, to very high levels of indoor air pollution. Partly as a result, respiratory illnesses are a major cause of death and illness in most LDCs.

AIR POLLUTION DAMAGE TO HUMAN HEALTH, PLANTS, AQUATIC LIFE, AND MATERIALS Your respiratory system has a number of mechanisms that help protect you from air pollution. For example, hairs in your nose filter out large particles. Sticky mucus in the lining of your upper respiratory tract captures small particles and dissolves some gaseous pollutants. Sneezing and coughing expel contaminated air and mucus when your respiratory system is irritated by pollutants. The cells of your upper respiratory tract are also lined with hundreds of thousands of tiny, mucus-coated hairlike structures called cilia. They continually wave back and forth, transporting mucus and the pollutants they trap to your throat, from which they are either swallowed or expelled.

Years of smoking and exposure to air pollutants can overload or break down these natural defenses. This causes or contributes to respiratory diseases such as *lung cancer*, *asthma* (typically an allergic reaction causing contraction of the bronchial tubes, resulting in acute shortness of breath), *chronic bronchitis* (damage to the cells lining the bronchial tubes, causing mucus buildup, coughing, and shortness of breath), and *emphysema* (damage to air sacs leading to abnormal dilation of air spaces, loss of lung elasticity, and acute shortness of breath). Elderly people, infants, pregnant women, and people with heart disease, asthma, or other respiratory diseases are especially vulnerable to air pollution.

No one knows how many people die prematurely from respiratory or cardiac problems caused or aggravated by air pollution. In the United States, estimates of annual deaths related to outdoor air pollution range from 7,000 to 180,000 people, and from 150,000 to 350,000 deaths occur if indoor air pollution is included. The wide range of these estimates shows how difficult it is to obtain accurate information concerning deaths attributable to air pollutants, mostly because of the large number of interacting factors affecting human health. According to the EPA and the American Lung Association, air pollution costs the United States at least $150 billion annually in health care and lost work productivity, with $100 billion of that caused by indoor air pollution.

The World Health Organization estimates that worldwide about 1.3 billion people—one person in four, mostly in LDCs—live in cities where outdoor air is unhealthy to breathe. The annual global death toll from air pollution is estimated to be at least four times the annual U.S. death toll from air pollution.

Some gaseous pollutants (especially ozone) damage leaves of crop plants and trees directly when they enter leaf pores. Spruce, fir, and other conifers, especially at high elevations, are most vulnerable to air pollution because of their long life spans and the year-round exposure of their needles to polluted air.

Prolonged exposure to high levels of multiple air pollutants can kill most trees and other vegetation in an area. However, the effects may not become visible for several decades, when large numbers of trees suddenly begin dying off because of depletion of soil nutrients and increased susceptibility to pests, diseases, fungi, and drought. This phenomenon, known as *Waldsterben* (forest death), has turned whole forests of spruce, fir, and beech in Europe into stump-studded meadows. A 1990 study estimated the cost of pollution damage to European forests to be roughly $30 billion per year.

Similar diebacks in the United States have occurred, mostly on high-elevation slopes that face moving air masses and are dominated by red spruce. Air pollution is also implicated in the recent dieback of sugar maples in Canada and the northeastern United States.

Air pollution, mostly by ozone, also threatens some crops—especially corn, wheat, soybeans, and peanuts—and is reducing U.S. food production by 5–10%. Estimates of economic losses to U.S. agriculture as a result of air pollution range from $1.9 billion to $5.4 billion per year.

High acidity can severely harm the aquatic life in freshwater lakes that have low alkaline content, or in areas in which surrounding soils have little acid-neu-

A: Renewable energy (92%), compared with coal (5%), oil (2.5%), and uranium (0.5%)

tralizing capacity. At least 16,000 lakes in Norway and Sweden contain no fish, and 52,000 more lakes have lost most of their acid-neutralizing capacity, because of excess acidity. In Canada, some 14,000 acidified lakes are almost fishless, and 150,000 more are in peril.

In the United States, about 9,000 lakes are threatened with excess acidity, one-third of them seriously. Most are concentrated in the Northeast and the upper Midwest—especially Minnesota, Wisconsin, and the upper Great Lakes—where 80% of the lakes and streams are threatened by excess acidity. Over 200 lakes in New York's Adirondack Mountains are too acidic to support fish.

Acidified lakes can be neutralized by treating them or the surrounding soil with large amounts of limestone, but liming is an expensive and only temporary remedy. Moreover, it can kill some types of plankton and aquatic plants, and it can harm wetland plants that need acidic water. It is also tricky to use correctly.

Each year air pollutants cause billions of dollars in damage to various materials we use. The fallout of soot and grit on buildings, cars, and clothing requires costly cleaning. Air pollutants break down exterior paint on cars and houses, and they deteriorate roofing materials. Irreplaceable marble statues, historic buildings, and stained-glass windows throughout the world have been pitted and discolored by air pollutants. Damage to buildings in the United States from acid deposition alone is estimated at $5 billion per year.

6-5 Solutions: Reducing Air Pollution

PREVENTING AND CONTROLLING OUTDOOR AIR POLLUTION Since 1970 the United States has significantly reduced levels of five of the seven major outdoor air pollutants. Even so, in 1992 at least 86 million people lived in areas that exceeded at least one air pollution standard. Emissions of nitrogen oxides have increased somewhat because of a combination of insufficient automobile emission standards and a growth in both the number of motor vehicles and the distances traveled. These conditions have also led to increases in ozone levels in many major urban areas.

A serious problem is that most U.S. air pollution control laws are based on pollution cleanup rather than pollution prevention. The only air pollutant with a sharp drop in its atmospheric level was lead, which was virtually banned in gasoline. This shows the effectiveness of the pollution prevention approach.

Figure 6-13 summarizes ways to reduce emissions of sulfur oxides, nitrogen oxides, and particulate matter from stationary sources such as electric power plants and industrial plants that burn coal. So far, most of the emphasis has been on dispersing the pollutants with tall smokestacks or adding equipment that removes some of the pollutants after they are produced. Environmentalists call for taxes on air pollutant emissions and a shift to prevention methods.

Figure 6-14 shows ways to reduce emissions from motor vehicles, the primary culprits in producing photochemical smog, which contains damaging ozone. Alternate fuels were evaluated in Table 5-1, p. 95.

Despite strict air pollution control laws, one of every four people in the United States is routinely exposed to ozone concentrations that exceed standards set under the Clean Air Act. However, progress is being made. From 1982 to 1992 overall U.S. smog incidence dropped by 8%, even as population and consumption rose. And during this same period the number of people exposed to unhealthful ozone levels in Los Angeles was cut in half. Even so, smog levels in Los Angeles are too high much of the year and could rise as population and consumption increase unless more stringent actions are taken to control air pollution in this area.

In 1989, California's South Coast Air Quality Management District Council proposed a drastic program to reduce ozone and photochemical smog in the Los Angeles area. This plan would do the following things:

- *Require 10% of new cars sold in California by 2003 to emit no air pollutants.*
- *Outlaw drive-through facilities to keep vehicles from idling in lines.*
- *Substantially raise parking fees and assess high fees for families owning more than one car.*
- *Strictly control or relocate industrial plants that release large quantities of hydrocarbons and other pollutants.* These facilities include petroleum-refining, dry-cleaning, auto-painting, printing, baking, and trash-burning plants.
- *Find substitutes for or ban use of consumer products that release hydrocarbons,* including aerosol propellants, paints, household cleaners, and barbecue starter fluids.
- *Eliminate gasoline-burning engines over two decades* by converting trucks, buses, chain saws, outboard motors, and lawn mowers to run on electricity or on alternative fuels (Table 5-1).
- *Require gas stations to use a hydrocarbon-vapor recovery system on gas pumps and to sell alternative fuels.*

Q: What percentage of the earth's proven reserves of oil are in OPEC countries?

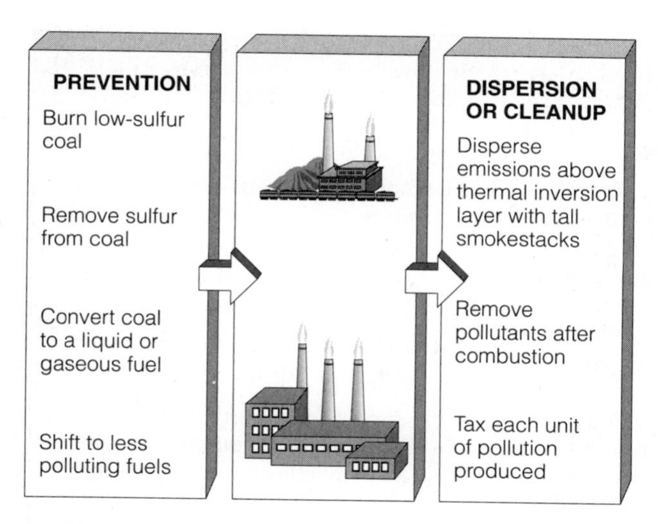

Figure 6-13 Methods for reducing emissions of sulfur oxides, nitrogen oxides, and particulate matter from stationary sources such as coal-burning electric power plants and industrial plants.

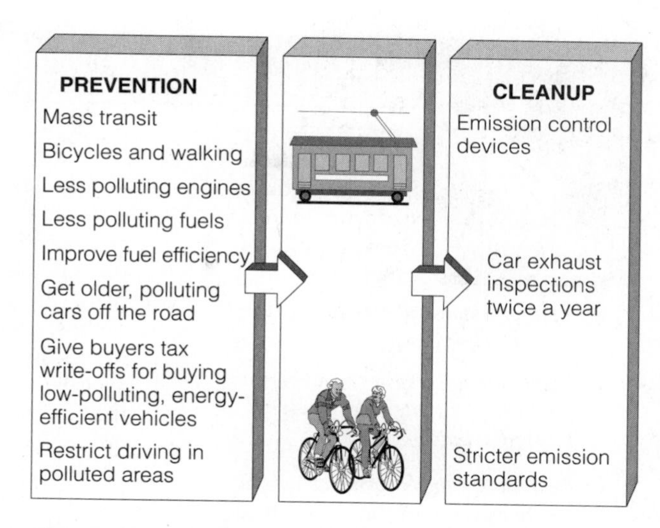

Figure 6-14 Methods for reducing emissions from motor vehicles.

Such measures are a glimpse of what most cities may have to do as people, cars, and industries proliferate.

SOLUTIONS: REDUCING INDOOR AIR POLLUTION For many people, indoor air pollution poses a greater threat to health than does outdoor air pollution. Yet the EPA spends $200 million per year trying to reduce outdoor air pollution and only $5 million a year on indoor air pollution.

Here are some ways air pollution experts have suggested for reducing indoor air pollution:

- *Modifying building codes to prevent radon infiltration, or requiring use of air-to-air heat exchangers or other devices to change indoor air at regular intervals*
- *Requiring exhaust hoods or vent pipes for appliances that burn natural gas or another fossil fuel*
- *Setting formaldehyde emission standards for building, furniture, and carpet materials*
- *Equipping work stations with adjustable fresh air inputs* (much like those for passengers on commercial aircraft)
- *Finding substitutes for potentially harmful chemicals in aerosols, cleaning compounds, paints, and other products used indoors*

Each of us can also take actions to reduce our own exposure to polluted indoor air (Individuals Matter, p. 128).

CONNECTIONS: PROTECTING THE ATMOSPHERE Environmentalists believe that protecting the atmosphere and thus the health of people and many other organisms will require the following significant changes throughout the world:

- *Integrate air pollution, water pollution, energy, land-use, and population policies.*
- *Emphasize pollution prevention rather than pollution control.* Widespread use of solar-produced hydrogen fuel (Section 5-5) would eliminate most air pollution.
- *Improve energy efficiency* (Section 5-2).
- *Reduce use of fossil fuels, especially coal and oil* (Section 5-7).
- *Shift to renewable energy resources* (Sections 5-3 and 5-4).
- *Emphasize distribution of low-emission and more efficiently vented cookstoves in rural areas of LDCs.*
- *Increase recycling and reuse, and reduce the production of all forms of waste* (Figure 1-14).
- *Slow population growth* (Section 2-4).
- *Include the external costs of air pollution and other forms of pollution in the market prices of goods and services* (Solutions, p. 5).

Some people have wondered whether there is intelligent life in other parts of the universe; others wonder whether there is intelligent life on Earth. If we can deal with the interconnected problems of possible global warming, ozone depletion, outdoor and indoor air pollution, loss of biodiversity (Chapter 4), and population growth (Chapter 2), then the answer is a hopeful yes.

A: 67% (25% in Saudi Arabia), compared to only 4% in the United States

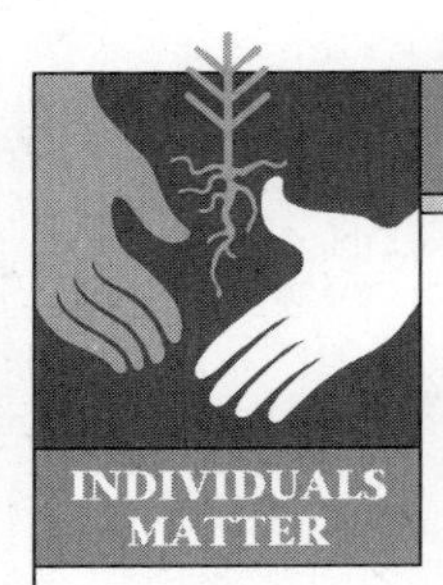

What *You* Can Do About Indoor Air Pollution

To reduce your exposure to indoor air pollutants:

- *Test for radon and take corrective measures as needed.*
- *Install air-to-air heat exchangers or regularly ventilate your house by opening windows.*
- *Test indoor air for formaldehyde* at the beginning of the winter heating season, when the house is closed up.* The cost is $200–$300.
- *Don't buy synthetic wall-to-wall carpeting, furniture, and other products containing formaldehyde, and use "low-emitting formaldehyde" or nonformaldehyde building materials.*
- *Reduce indoor levels of formaldehyde and several other toxic gases by growing house plants.* Examples are the spider or airplane plant (the most effective), golden pothos, syngonium, philodendron (especially the elephant-ear species), chrysanthemum, ligustrum, photina, variegated liriope, and Gerbera daisy. About 20 plants of such species can help clean the air in a typical home. Plants should be potted with a mixture of soil and granular charcoal (which absorbs organic air pollutants).
- *Test your house or workplace for asbestos fiber levels if it was built before 1980.*† Don't buy a pre-1980 house without having its indoor air tested for asbestos.
- *Don't store gasoline, solvents, or other volatile hazardous chemicals inside a home or attached garage.*
- *Don't use commercial room deodorizers or air fresheners.*
- *Don't use aerosol spray products.*
- *Don't smoke.* If you must smoke, do it outside or in a closed room vented to the outside.
- *Make sure that wood-burning stoves, fireplaces, and kerosene- and gas-burning heaters are properly installed, vented, and maintained.*

* To locate a testing laboratory in your area, write to Consumer Product Safety Commission, Washington, DC 20207, or call 301-492-6800.

† To get a free list of certified asbestos laboratories that charge $25–$50 to test a sample, send a self addressed envelope to NIST/NVLAP, Building 411, Room A124, Gaithersburg, MD 20899, or call the EPA's Toxic Substances Control Hotline at 202-554-1404.

The atmosphere is the key symbol of global interdependence. If we can't solve some of our problems in the face of threats to this global commons, then I can't be very optimistic about the future of the world.

MARGARET MEAD

CRITICAL THINKING

1. What consumption patterns and other features of your lifestyle directly add greenhouse gases to the atmosphere? Which, if any, of those things would you be willing to give up to slow possible global warming and reduce other forms of air pollution?
2. Explain why you agree or disagree with each of the proposals listed in Section 6-2 for **(a)** slowing down emissions of greenhouse gases into the atmosphere and **(b)** preparing for the effects of global warming.
3. What consumption patterns and other features of your lifestyle directly and indirectly add ozone-depleting chemicals to the atmosphere? Which, if any, of those things would you be willing to give up to slow ozone depletion?
4. Should all tall smokestacks be banned? Explain your answers.
5. Explain why you agree or disagree with each of the proposals made by the American Medical Association (listed on p. 124) for reducing the harmful effects of smoking.

Q: How much of the oil used in the United States is imported?

7 Water

Our liquid planet glows like a soft blue sapphire in the hard-edged darkness of space. There is nothing else like it in the solar system. It is because of water.

JOHN TODD

7-1 Water's Importance and Unique Properties

We live on the water planet. A precious film of water—most of it salt water—covers about 71% of the earth's surface. Earth's organisms are made up mostly of water. For example, a tree is about 60% water by weight, and you and most animals are about 65% water.

Fresh water is a vital resource for agriculture, manufacturing, transportation, and countless other human activities. Water also plays a key role in sculpting the earth's surface, moderating climate, and diluting pollutants.

Water has many unique—almost magical—properties. Its high boiling point and low freezing point mean that water remains a liquid in most climates on Earth (Connections, p. 109). It can store a large amount of heat without a large change in temperature, which helps protect living organisms from the shock of abrupt temperature changes, moderates the earth's climate, and makes water an excellent coolant. Water's ability to absorb large amounts of heat as it changes into water vapor—and to release this heat as the vapor condenses back to liquid water—is a primary factor in distributing heat throughout the world.

Water can also dissolve a variety of compounds. This property enables it to carry dissolved nutrients into the tissues of living organisms, to flush waste products out of those tissues, to serve as an all-purpose cleanser, and to help remove and dilute the water-soluble wastes of civilization. However, water's superiority as a solvent also means that it is easily polluted by water-soluble wastes.

Most substances shrink when they freeze, but liquid water expands when it becomes ice. Consequently, ice has a lower density (mass per unit of volume) than liquid water, and thus ice floats on water, and bodies of water freeze from the top down instead of from the bottom up. Without this property, lakes and streams in cold climates would freeze solid, and most current forms of aquatic life could not exist.

Water is truly a wondrous substance that connects us to one another, to other forms of life, and to the entire planet. Despite its importance, water is one of the earth's most poorly managed resources. We waste it and pollute it; we also charge too little for making it available, thus encouraging even greater waste and pollution of this vital renewable resource.

7-2 Supply, Renewal, and Use of Water Resources

THE WATER CYCLE Only a tiny fraction of the planet's abundant water is available to us as fresh water. About 97% is found in the oceans and is too salty for drinking, irrigation, or industry (except as a coolant). The remaining 3% is fresh water. About 2.997% of it is locked up in ice caps or glaciers, or is buried so deep that it costs too much to extract. Only about 0.003% of the earth's total volume of water is easily available to us as soil moisture, exploitable groundwater, water vapor, and lakes and streams. If the world's water supply were only 100 liters (26 gallons), our usable supply of fresh water would be only about 0.003 liter (one-half teaspoon).

Fortunately, the available fresh water amounts to a generous supply that is continuously collected, purified, and recycled in the *hydrologic cycle*. The fresh water we use comes from surface water and groundwater, which become available from precipitation (Figure 7-1). Water in turn goes from these sources into the atmosphere via evaporation and transpiration.

This natural recycling and purification process provides plenty of fresh water so long as we don't overload it with slowly degradable and nondegradable wastes or withdraw water from underground supplies faster than it is replenished. Unfortunately

A: 51% in 1993

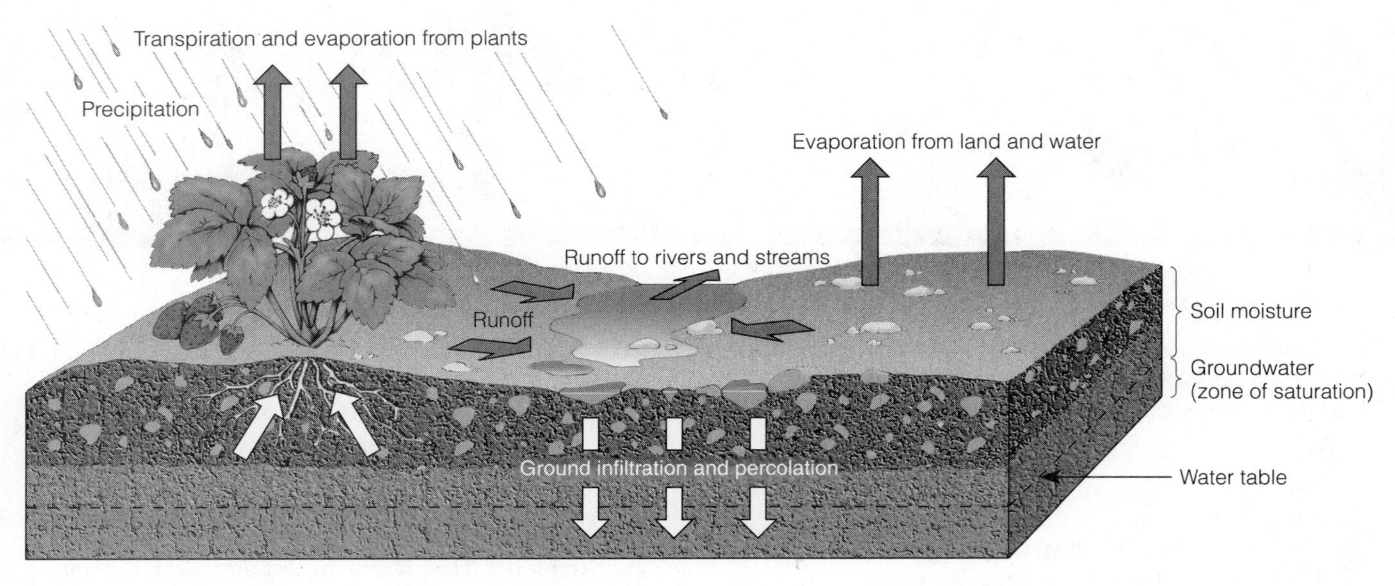

Figure 7-1 Main routes of local precipitation: surface runoff into surface waters, ground infiltration into aquifers, and evaporation and transpiration into the atmosphere.

we are doing both. Further, usable fresh water is unevenly distributed around the world. Differences in average annual precipitation divide the world into water "haves" and "have-nots."

As population and industrialization increase, water shortages in already water-short regions are expected to intensify. Projected global warming (Section 6-2) also might cause changes in rainfall patterns and disrupt water supplies, and we can't predict which areas might be affected.

THE IMPORTANCE OF OCEANS A more accurate name for Planet Earth would be Ocean, because oceans cover about 71% of its surface. By serving as a gigantic reservoir for carbon dioxide, oceans help regulate the temperature of the troposphere and thus global climate. Oceans provide habitats for about 250,000 species of marine plants and animals, which are food for many other organisms, including human beings; they also serve as a source of iron, sand, gravel, phosphates, magnesium, oil, natural gas, and many other valuable resources. Because of their size and currents, oceans absorb many human-produced wastes that flow or are dumped into them and mix and dilute them to less harmful or even harmless levels, so long as the oceans are not overloaded.

Oceans have two major life zones: the coastal zone and the open sea (Figure 7-2). Although it makes up less than 10% of the ocean's surface area, the *coastal zone* contains 90% of all marine species and is where most of the fish and shellfish we eat are caught. Highly productive ecosystems in the coastal zone include the following:

- *Coral reefs*—often found in warm tropical and subtropical oceans. They are formed by massive colonies consisting of billions of tiny coral animals called polyps. Coral reefs provide habitats for a great diversity of marine life and help protect 15% of the world's coastlines from storms by reducing the energy of incoming waves. These ecosystems grow slowly and are easily disrupted.
- *Estuaries*—coastal areas at the mouths of rivers, whose fresh water, carrying fertile silt and runoff from the land, mixes with salty seawater.
- *Coastal wetlands*—land in a coastal area covered all or part of the year with salt water. They are breeding grounds and habitats for waterfowl and other wildlife and dilute and filter out large amounts of nutrients and waterborne pollutants. In temperate areas these wetlands usually consist of a mix of bays, lagoons, salt flats, mud flats, and salt marshes in which grasses are the dominant vegetation. In warm tropical climates we find highly productive mangrove swamps dominated by mangrove trees, any of about 55 species of trees and shrubs that can live partly submerged in the relatively salty environment of coastal swamps and that help protect coastlines from storm damage.
- *Barrier islands*—long, thin, low, offshore islands of sediment that generally run parallel to the shore along some coasts (such as most of North America's Atlantic and Gulf coasts). These islands help protect the mainland, estuaries, lagoons, and coastal wetlands by dispersing the energy of approaching storm waves.

 Q: Adjusting for inflation, how does the price of gasoline in the United States in 1993 compare with its price in 1950? In 1973?

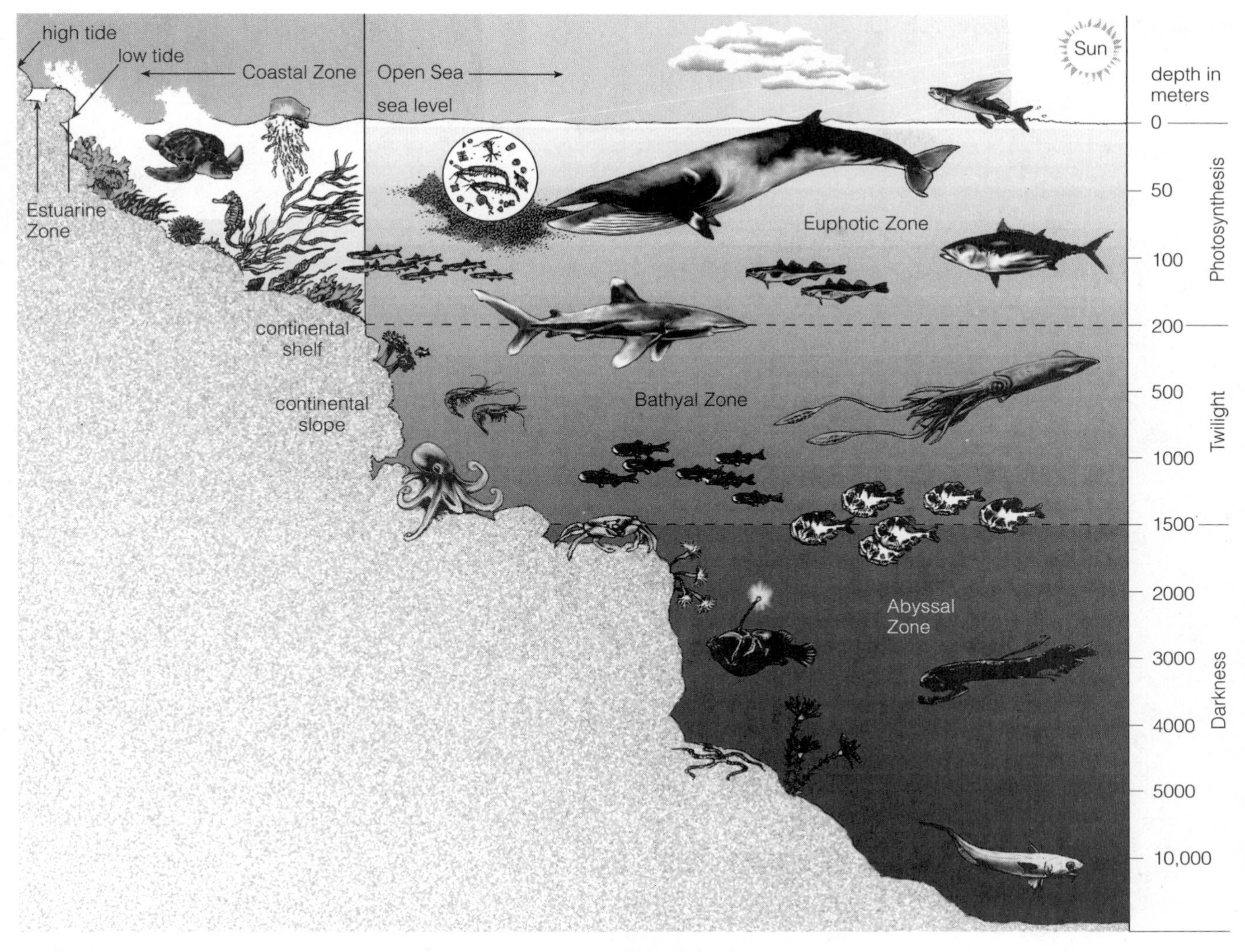

Figure 7-2 Major life zones in an ocean (actual depths of zones may vary). About 97% of the earth's water is in the interconnected oceans. The average depth of the world's oceans is 3.8 kilometers (2.4 miles).

Unfortunately, these important coastal ecosystems are under severe stress from human activities.

The *open sea* is divided into three vertical zones—euphotic, bathyal, and abyssal—based primarily on the penetration of sunlight (Figure 7-2). This vast volume of ocean contains only about 10% of all marine species.

FRESHWATER LAKES AND STREAMS *Lakes* are large natural bodies of standing fresh water formed when precipitation, surface runoff, or groundwater flowing from underground springs fills depressions in the earth. Lakes in temperate zones normally consist of distinct zones (Figure 7-3), providing habitats and niches for various species.

Precipitation that doesn't sink into the ground or evaporate becomes **runoff** (Figure 7-1), which flows into streams and eventually to the ocean to continue circulating in the hydrologic cycle. The entire land area that delivers water, sediment, and dissolved substances via small streams to a larger stream (or river), and ultimately to the sea, is called a **watershed**, or a **drainage basin**.

The downward flow of water from mountain highlands to the sea takes place in three zones in a *river system* (Figure 7-4, p. 133). Because of different environmental conditions in each zone, a river system is a series of different ecosystems.

INLAND WETLANDS Lands covered with fresh water at least part of the year (excluding lakes, reservoirs, and streams) and located away from coastal areas are called **inland wetlands**. They include marsh

A: About the same in both cases

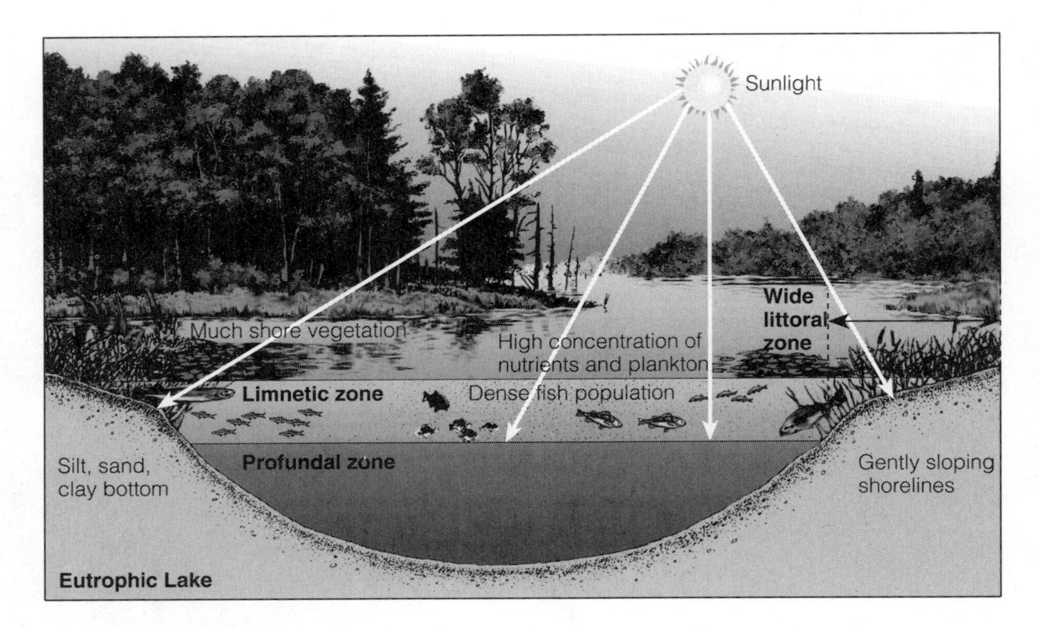

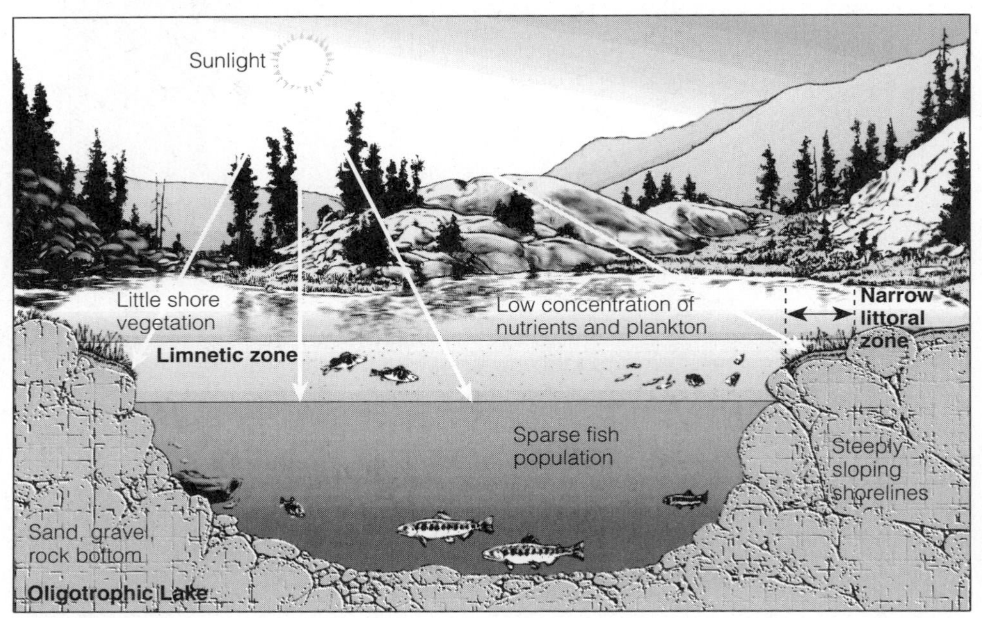

Figure 7-3 The distinct zones of life in a temperate zone lake.

bogs, prairie potholes, swamps, mud flats, floodplains, fens, wet meadows, and the wet arctic tundra in summer.

Some wetlands are covered with water year-round. Others, such as prairie potholes, floodplain wetlands, and bottomland hardwood swamps are *seasonal wetlands* that are underwater or soggy for only a short time each year.

Inland wetlands provide habitats for fish, waterfowl, and other wildlife, and they improve water quality by filtering, diluting, and degrading sediments and pollutants as water flows through. Floodplain wetlands near rivers help regulate stream flow by storing water during periods of heavy rainfall and releasing it slowly, which reduces riverbank erosion and flood damage.

Despite the ecological importance of year-round and seasonal inland wetlands, they are under attack. Each year some 1,200 square kilometers (470 square miles) of inland wetland in the United States are lost, about 80% to agriculture and the rest to mining, forestry, oil and gas extraction, highways, and urban development. Many other countries are experiencing similar losses.

GROUNDWATER Some precipitation infiltrates the ground and fills the pores in soil and rock. The subsurface area where all available soil and rock spaces

Q: How long will the world's oil reserves last at the current consumption rate?

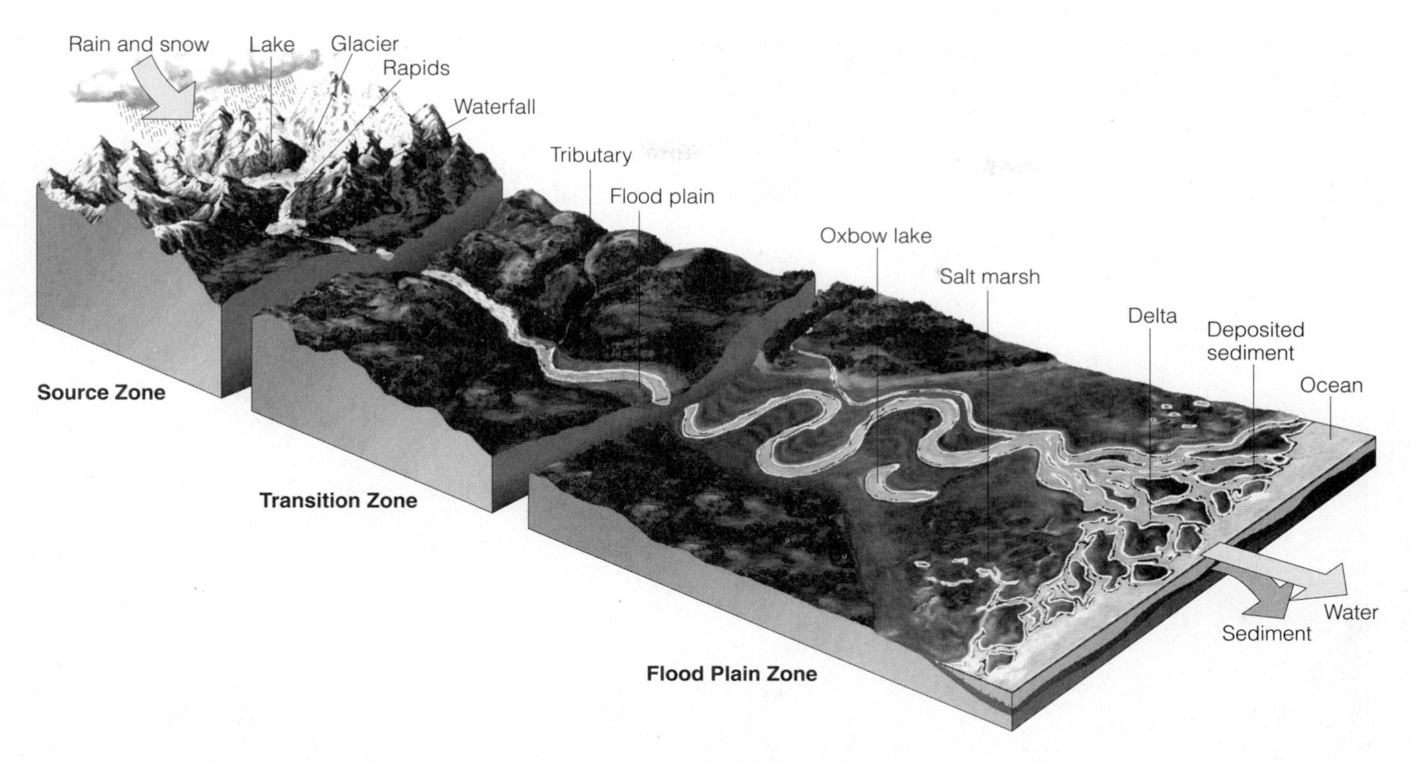

Figure 7-4 The three zones in the flow of water downhill from mountain (headwater) streams to wider, lower-elevation streams to rivers, which empty into the ocean.

are filled by water is called the **zone of saturation**, and the water in these pores is called **groundwater** (Figure 7-1). The **water table**, which is the upper surface of the zone of saturation, falls in dry weather and rises in wet weather. Porous, water-saturated layers of sand, gravel, or bedrock through which groundwater flows (and that can yield an economically significant amount of water) are called **aquifers** (Figure 7-5).

Most aquifers are replenished naturally by precipitation, which percolates downward through soil and rock in what is called **natural recharge**. Normally, groundwater moves from points of high elevation and pressure to points of lower elevation and pressure. This movement is quite slow, typically only a meter or so (about 3 feet) per year and rarely more than 0.3 meter (1 foot) per day. Thus, most aquifers are like huge, slow-moving underground lakes.

If the withdrawal rate of an aquifer exceeds its natural recharge rate, the water table around the withdrawal well is lowered, creating a waterless volume known as a *cone of depression*. Any pollutant discharged onto the land above will be pulled directly into this cone and will pollute water withdrawn by the well.

Some aquifers, called *fossil aquifers*, get very little—if any—recharge. Often found deep underground, they are (on a human time scale) nonrenewable resources. Withdrawals from fossil aquifers amount to "water mining" that, if kept up, will deplete these ancient deposits of liquid Earth capital.

GLOBAL WATER USE Since 1950 the global rate of water withdrawn from groundwater or surface-water sources for human use has increased almost fivefold, and per capita use has trebled, largely to meet the food and other resource needs of the world's rapidly growing population. Water withdrawal rates are projected to at least double in the next two decades.

Averaged globally, about 69% of the water withdrawn each year is used to irrigate 18% of the world's cropland, much of it in the United States, the former Soviet Union, and Mexico. A large share of irrigation water is wasted, with 70–80% of the water either evaporating or seeping into the ground before reaching crops.

Worldwide, about 23% of the water withdrawn is used for energy production (oil and gas production and power-plant cooling) and industrial processing, cleaning, and waste removal. Domestic and municipal use accounts for about 8% of worldwide withdrawals (13–16% in MDCs). As population, urbanization, and industrialization grow, the volume of wastewater needing treatment will increase enormously.

A: 42 years (undiscovered oil might add another 40 years)

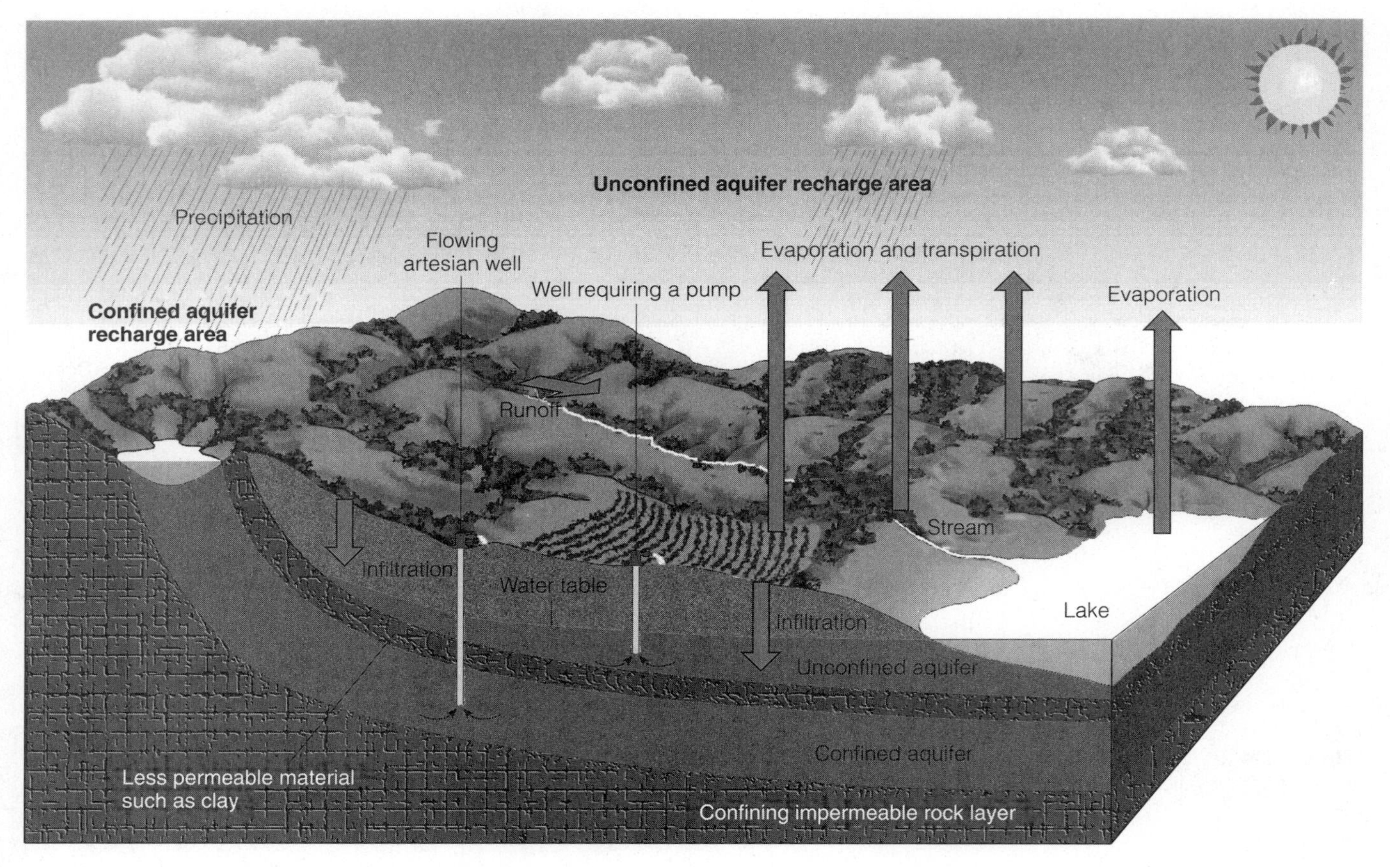

Figure 7-5 The groundwater system. An *unconfined (water table) aquifer* forms when groundwater collects above a layer of rock or compacted clay through which water flows very slowly (low permeability). A *confined aquifer* is sandwiched between layers such as clay or shale that have low permeability. Groundwater in this type of aquifer is confined and under pressure.

7-3 Water Resources Problems and Solutions

TOO LITTLE FRESH WATER *Droughts*—periods in which precipitation is much lower and evaporation is higher than normal—cause more damage and suffering worldwide than any other natural hazard. Since the 1970s drought has killed more than 24,000 people per year and created swarms of environmental refugees. At least 80 arid and semiarid countries, where nearly 40% of the world's people live, experience years-long droughts.

Reduced precipitation, higher-than-normal temperatures, or both usually trigger a drought; rapid population growth makes it worse. Deforestation (Sections 4-2 and 4-3), overgrazing by livestock (Section 4-4), desertification (Figure 3-7), and replacing diverse natural grasslands with monoculture fields of crops can intensify the effects of drought.

Millions of poor people in LDCs have no choice but to try to survive on drought-prone land. In water-short LDCs many women and children must walk long distances each day, carrying heavy jars or cans, to get a meager supply of sometimes contaminated water.

If global warming occurs as projected (Section 6-2), severe droughts may become more frequent in some areas of the world and may jeopardize food production. Some water-starved cities may have to be abandoned. Competition between cities and farmers for scarce water is already escalating in the western United States and China.

Water will be the burning foreign policy issue for water-short countries in the 1990s and beyond. Almost 150 of the world's 214 major river systems are shared by two countries, and another 50 are shared by three or more nations. Thus 40% of the world's population already clashes over water, especially in the mostly arid Middle East. The next wars in the Middle East may well be fought over water, not oil.

Q: How much new oil must be discovered and developed to continue using oil at the current rate?

Some countries have lots of water, but the largest rivers (which carry most of the runoff) are far from agricultural and population centers where the water is needed. For example, South America has the largest annual water runoff of any continent, but 60% of the runoff flows in the Amazon River through remote areas where few people live.

LDCs rarely have the money to develop the water storage and distribution systems needed to increase their supply; their people must settle where the water is. In MDCs people tend to live where the climate is favorable and bring in water from another watershed.

Strategies for capturing some of this runoff and bringing the water to people include building dams and reservoirs (Figure 5-11) and using tunnels, aqueducts, and pipes to transport water to other areas. These approaches, however, are expensive and have harmful environmental impacts in addition to their benefits (Figure 5-11 and Connections, right).

Desalination—the removal of dissolved salts from ocean water or from brackish (slightly salty) groundwater—is another way to increase fresh water supplies. Currently about 7,500 desalination plants in 120 countries provide about 0.1% of the fresh water used by humans. Desalination, however, uses vast amounts of electricity, and thus desalinated water costs three to five times more than water from conventional sources.

Desalination can provide fresh water for coastal cities in arid countries, such as sparsely populated Saudi Arabia, where the cost of getting fresh water by any method is high. However, desalinated water will probably never be cheap enough to use for irrigating conventional crops or to meet much of the world's demand for fresh water, unless efficient solar-powered methods can be developed.

TOO MUCH WATER Some countries have enough annual precipitation but get most of it at one time of the year. In India, for example, 90% of the annual precipitation falls between June and September, the monsoon season. This prolonged downpour causes floods, waterlogs soils, leaches soil nutrients, and washes away topsoil and crops.

Natural flooding by streams, the most common type of flooding, is caused primarily by heavy rain or rapid melting of snow; this causes water in the stream to overflow the channel in which it normally flows and to cover the adjacent area, called a **floodplain**.

People have settled on floodplains since the beginnings of agriculture. The soil is fertile, and ample water is available for irrigation. Communities have access to the water for transportation of people or goods, and floodplains are flat sites, suitable for buildings, highways, and railroads. Prolonged rains can

CONNECTIONS

The Aral Sea Ecological Disaster

One example of a massive water transfer project is diversion of water from rivers feeding the inland Aral Sea (a huge freshwater lake in Central Asia in the former Soviet Union) to irrigate cropland. The diversion (coupled with droughts) has caused a regional ecological disaster, described by one former Soviet official as "ten times worse than the 1986 Chernobyl nuclear power plant accident." The sea's salinity has tripled, its surface area has shrunk by 46%, and its volume has decreased by 69%.

All the native fish are gone, devastating the area's fishing industry, which once provided work for more than 60,000 people. Roughly half of the area's bird and mammal species have also disappeared. Salt, dust, and dried pesticide residues have been carried as far as 300 kilometers (190 miles) by the wind. As the salt spreads, it kills crops, trees, and wildlife and destroys pastureland.

The once-huge Aral Sea acted as a thermal buffer, moderating the heat of summer and the extreme cold of winter. Now there is less rain, summers are hotter, winters are colder, and the growing season is shorter. Cotton and crop yields have dropped dramatically.

Local farmers have turned to using pesticides and fertilizers on some crops. Many of these chemicals have percolated downward and accumulated to dangerous levels in the groundwater, from which most of the drinking water comes.

Ways to deal with this problem include **(1)** charging farmers more for irrigation water to reduce waste and encourage a shift to less water-intensive crops; **(2)** decreasing irrigation water quotas; **(3)** introducing water-saving technologies; **(4)** developing a regional integrated water management plan; **(5)** planting protective forest belts; **(6)** using underground water to supplement irrigation water and lower the water table to reduce waterlogging and salinization; **(7)** improving health services; and **(8)** slowing the area's rapid population growth (3% per year).

In 1992, Kazakhstan, Uzbekistan, Kyrgyzstan, and Turkmenistan, which share the Aral Sea basin, signed an agreement describing how the waters of the two rivers feeding the Aral Sea should be shared, and they created a council to manage the basin's resources. However, even with help from foreign countries, the United Nations, and agencies such as the World Bank, the money needed to save the Aral Sea may not be available.

A: A supply equal to the reserve in Saudi Arabia (the world's largest) every 10 years

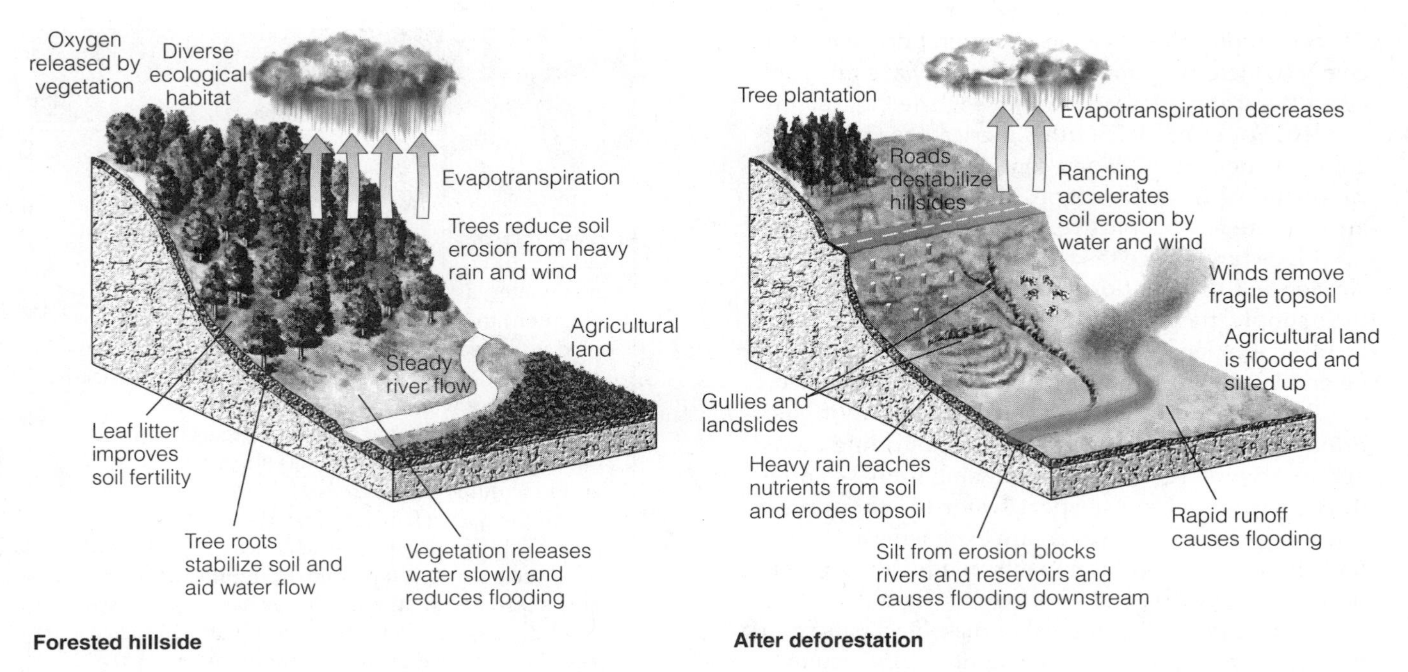

Figure 7-6 A mountainside before and after deforestation. When a hillside or mountainside is deforested—for timber and fuelwood, for grazing livestock, or for unsustainable farming—water from rains rushes down denuded slopes, eroding precious topsoil and flooding downstream areas. A 3,000-year-old Chinese proverb says: "To protect your rivers, protect your mountains."

cause streams and lakes anywhere to overflow and flood the surrounding floodplain, but low-lying river basins such as the Ganges River basin in India and Bangladesh are especially vulnerable. Hurricanes and typhoons can also flood low-lying coastal areas.

In the 1970s floods killed more than 4,700 people per year and caused tens of billions of dollars in property damage, a trend that continued (and even worsened) in the 1980s and early 1990s, including massive flooding in the midwestern United States in 1993.

Floods, like droughts, are usually considered to be natural disasters, but human activities have contributed to the sharp rise in flood deaths and damages since the 1960s. The main way humans increase the probability and severity of flooding is by removing vegetation—through logging (Figure 7-6), overgrazing by livestock, construction, forest fires, and certain mining activities. Vegetation retards surface runoff and increases infiltration; when the vegetation is removed by human activities or natural occurrences, precipitation reaches streams more directly, often with a large load of sediment, which increases the chance of flooding by making the stream more shallow.

Urbanization also increases flooding (even with moderate rainfall) by replacing vegetation and soil with highways, parking lots, and buildings that lead to rapid runoff of rainwater. If sea levels rise during the next century, as projected because of global warming (Section 6-2), many low-lying coastal cities, wetlands, and croplands will be under water.

Ways to reduce the risks of flooding include the following:

- *Deepening, widening, or straightening a section of a stream (channelization)* to allow more rapid runoff and reduce the chances of flooding. However, the increased flow of water can increase both upstream erosion and downstream deposits of sediment.
- *Building artificial levees along stream banks* to reduce the chances of water flowing over into nearby floodplains. However, even the best-designed levees can be breached by flooding from prolonged rainfall, as many people in the midwestern United States learned during the summer of 1993.
- *Building a flood control dam across a stream* to hold back and store flood water and release it more gradually. This has both benefits and drawbacks (Figure 5-11).
- *Managing floodplains*. Historical data on flood frequency and severity are used to prohibit certain types of buildings or activities in high-risk areas, to elevate or allow only flood-proof buildings on the legally defined floodplain, and in some cases to construct a floodway that allows flood water to flow through the community with

 Q: What would happen if oil's harmful effects were included in its market price and government subsidies were removed?

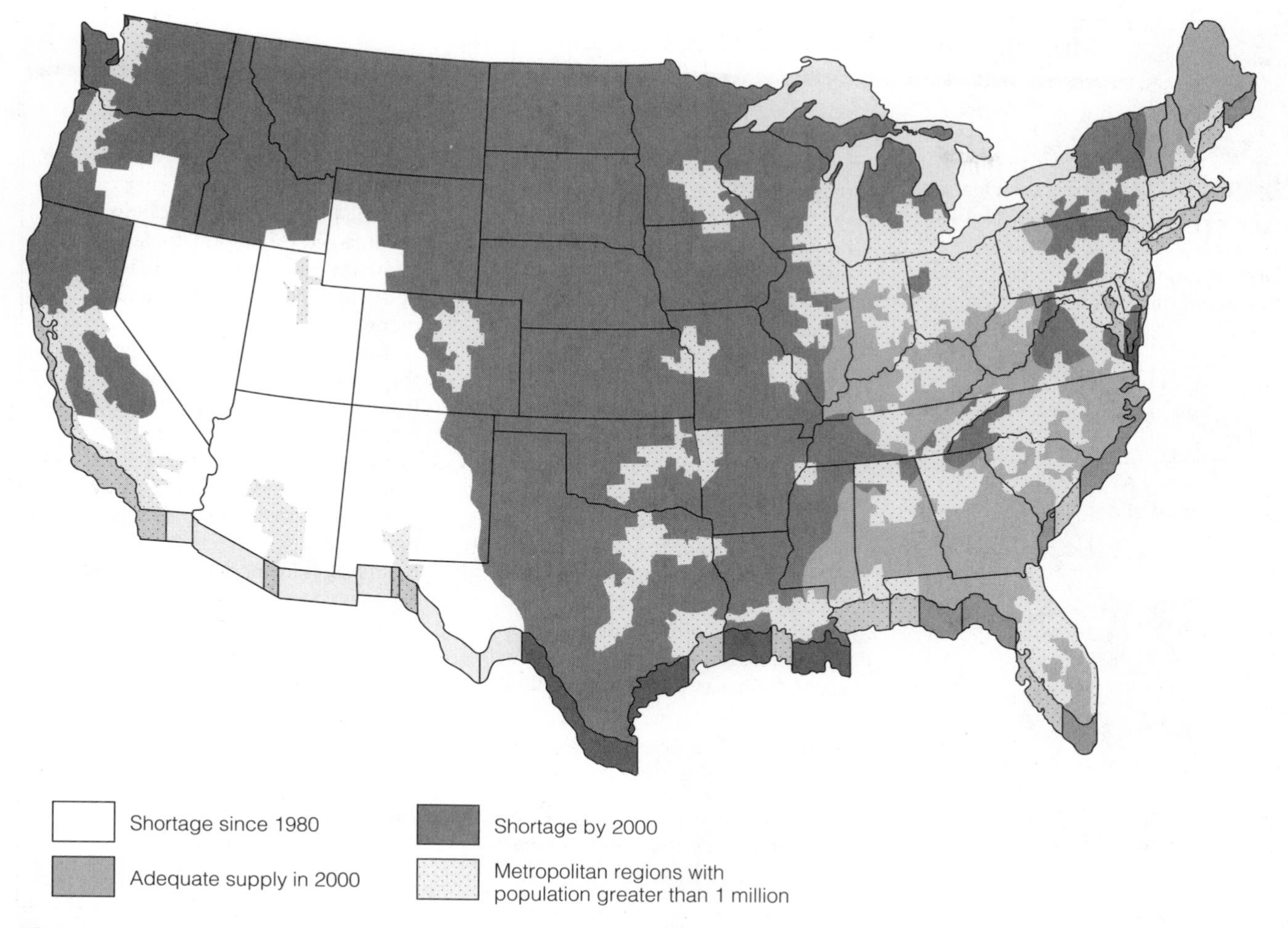

Figure 7-7 Current and projected water-deficit regions in the continental United States and their proximity to metropolitan areas having populations greater than 1 million. (Data from U.S. Water Resources Council and U.S. Geological Survey)

minimal damage. Most experts consider this the best way to minimize flood damage.

WATER RESOURCES IN THE UNITED STATES

Although the United States has plenty of fresh water, much of it is in the wrong place at the wrong time or is contaminated by agriculture and industry. From a human perspective, the eastern states usually have ample precipitation, whereas many of the western states have too little. In the East the largest uses for water are for energy production, cooling, and manufacturing; in the West the largest use by far is for irrigation.

Many major urban centers, especially those in the West and Midwest, are located in areas that don't have enough water or are projected to have water shortages by 2000 (Figure 7-7). Experts project that current shortages and conflicts over water supplies will get much worse as more industries and people migrate west and compete with farmers for scarce water. These shortages could worsen even more if climate warms as a result of an enhanced greenhouse effect (Section 6-2). Because water is such a vital resource, the data depicted in Figure 7-7 may be useful in deciding where to live in the coming decades.

GROUNDWATER DEPLETION In the United States, 23% of all fresh water used is groundwater. About half of the country's drinking water (96% in rural areas and 20% in urban areas) and 40% of irrigation water are pumped from aquifers.

Overuse of groundwater can cause or intensify *aquifer depletion*, *aquifer subsidence* (sinking of land when groundwater is withdrawn), and *intrusion of salt water into aquifers*. Groundwater can also become contaminated from industrial and agricultural activities, septic tanks, and other sources, as discussed in Section 7-4.

Currently, about one-fourth of the groundwater withdrawn in the United States is not replenished. The most serious overdraft is in parts of the huge Ogallala

A: It would be too expensive to use and would be phased out

Mining Water—The Shrinking Ogallala Aquifer

CASE STUDY

Water pumped from the Ogallala Aquifer (Figure 7-8)—the world's largest known aquifer—has helped transform much of a vast prairie that is too dry for rainfall farming into one of the United States's most productive farmlands. Although this aquifer is gigantic, it is essentially a nonrenewable fossil aquifer with an extremely slow recharge rate. Water is being pumped out eight times as fast as natural recharge occurs, mostly for irrigation to supply 15% of the country's corn and wheat, 25% of its cotton, and 40% of its beef.

The withdrawal rate is 100 times the recharge rate for parts of the aquifer that lie beneath Texas, New Mexico, Oklahoma, and Colorado. Water experts project that at the current rate of withdrawal, one-fourth of the aquifer's original supply will be depleted by 2020—much sooner in areas where it is shallow; then the area will become a desert. It will take thousands of years to replenish the aquifer. Depletion is encouraged by federal tax laws that allow farmers and ranchers to deduct the cost of drilling equipment and sinking wells.

Long before the water is gone, the high cost of pumping water from a rapidly dropping water table will force many farmers to grow drought-tolerant crops instead of profitable but "thirsty" crops such as cotton and sugar beets. Some farmers will go out of business. If farmers in the Ogallala region conserve more water and switch to crops that require less water, depletion of the aquifer could be delayed.

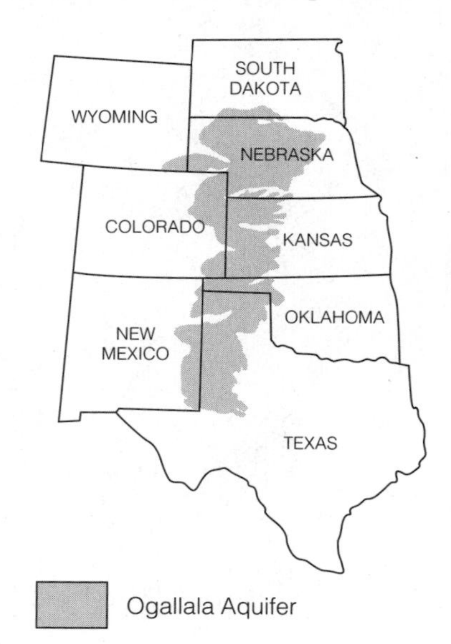

Figure 7-8 The Ogallala, the world's largest known aquifer. If the water in this aquifer were above ground, it would be enough to cover the entire lower 48 states with 0.5 meter (1.5 feet) of water. This fossil aquifer, which is renewed very slowly, is being depleted to grow crops and raise cattle.

Aquifer, extending from southern South Dakota to northwestern Texas (Case Study, above). Aquifer depletion is also a serious problem in Saudi Arabia, northern China, Mexico, Thailand, and parts of India.

Ways to slow groundwater depletion include **(1)** controlling population growth, **(2)** discontinuing planting of water-thirsty crops in dry areas, **(3)** developing crop strains that require less water and are more resistant to heat stress, and **(4)** wasting less irrigation water.

REDUCING WASTE OF WATER RESOURCES

Mohamed El-Ashry of the World Resources Institute estimates that *65–70% of the water people use throughout the world is wasted through evaporation, leaks, and other losses*. The United States—the world's largest user of water—does slightly better but still wastes 50% of the water it withdraws. El-Ashry believes that it is economically and technically feasible to reduce water waste to 15%, thus meeting most of the world's water needs for the foreseeable future.

Conserving water would have many other benefits, including reducing the burden on wastewater plants and septic systems, decreasing pollution of surface water and groundwater, reducing the number of expensive dams and water-transfer projects that destroy wildlife habitats and displace people, slowing depletion of groundwater aquifers, and saving energy and money needed to supply and treat water.

A prime cause of water waste in the United States (and in most countries) is artificially low water prices. Cheap water is the main reason that farmers in Arizona and southern California can grow water-thirsty crops like alfalfa in the middle of a desert. It also enables people in Palm Springs, California, to keep their lawns and 74 golf courses green in a desert area.

Water subsidies are paid for by all taxpayers through higher taxes. Because these external costs don't show up on monthly water bills, consumers have little incentive to use less water or to install water-conserving devices and processes. Raising the price of water to reflect its true cost (Solutions, p. 5) would be a powerful incentive for using water more efficiently.

Q: How long will proven reserves of natural gas last at current consumption rates?

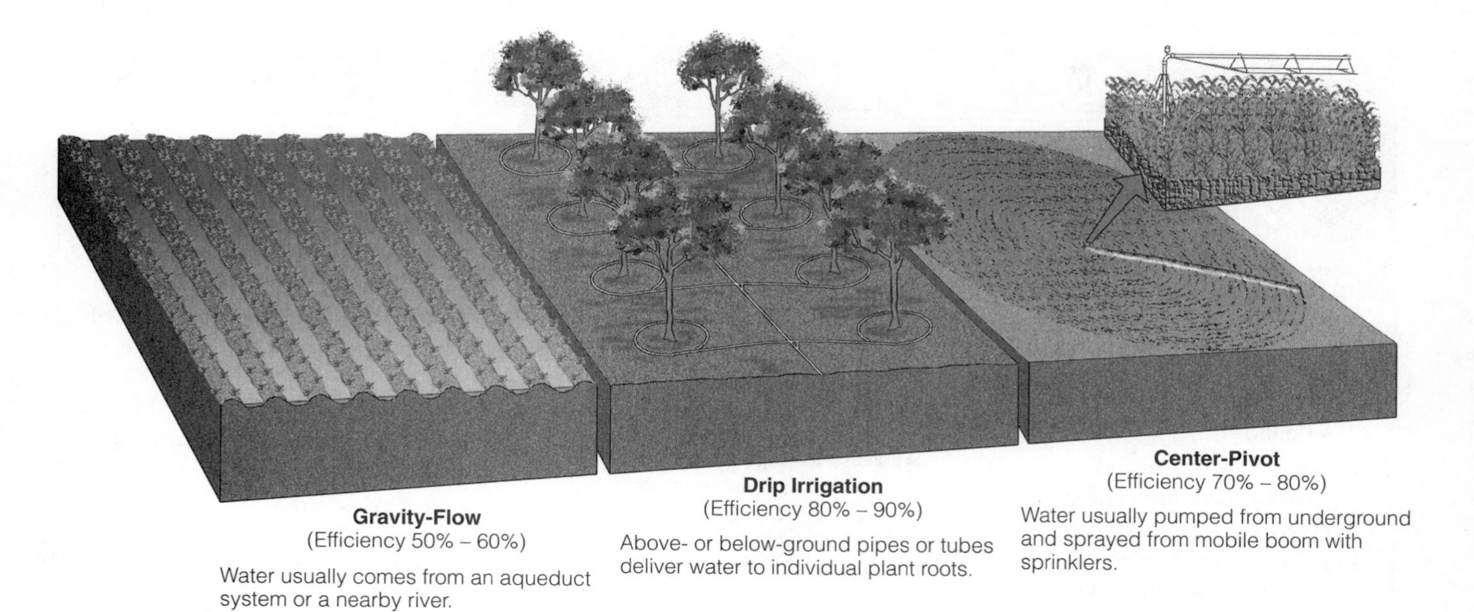

Figure 7-9 Major irrigation systems.

Water waste from irrigation (which accounts for 69% of global water withdrawal) can be reduced by **(1)** using more efficient drip and center-pivot irrigation (Figure 7-9), **(2)** lining irrigation ditches to prevent seepage, **(3)** using holding ponds to store rainfall or capture irrigation water for recycling to crops, **(4)** using computer-controlled systems that monitor soil moisture and provide water only when necessary, **(5)** growing crops with water-saving organic farming techniques, **(6)** switching to more drought-resistant and salt-tolerant crop varieties, and **(7)** using nutrient-rich, treated urban wastewater for irrigation.

Since 1950 water-short Israel has used many of these techniques to slash irrigation water waste by about 84% while irrigating 44% more land. However, so long as irrigation water is cheap and plentiful, farmers have little incentive to invest in water-saving techniques.

Manufacturing processes can use recycled water or be redesigned to save water. Japan and Israel lead the world in conserving and recycling water in industry. For example, a paper mill in Hadera, Israel, uses one-tenth as much water as most other paper mills do. Manufacturing aluminum from recycled scrap rather than from virgin ores can reduce water needs by 97%.

In the United States, industry is the largest conserver of water. However, the potential for water recycling in U.S. manufacturing has hardly been tapped because the cost of water to many industries is subsidized. Finally, each of us can reduce our own use and waste of water and in the process can often save money (Individuals Matter, p. 140).

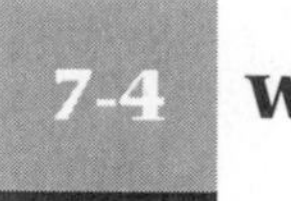

7-4 Water Pollution

CONTAMINATED DRINKING WATER In many parts of the world, water quality has been degraded. Rivers in eastern Europe, Latin America, and Asia are severely polluted, as are some in MDCs. Aquifers used as sources of drinking water in many MDCs and LDCs are becoming contaminated with pesticides, fertilizers, and hazardous organic chemicals. According to the World Health Organization, 1.5 billion people don't have a safe supply of drinking water, and 1.7 billion people lack adequate sanitation facilities. At least 5 million people die every year from waterborne diseases that could be prevented by clean drinking water and better sanitation. Most of the 13,700 who die each day from such diseases are children under age 5.

In 1980 the United Nations called for spending $300 billion to supply all of the world's people with clean drinking water and adequate sanitation by 1990. The annual $30-billion cost of this program is about what the world spends every 10 days for military purposes. Sadly, only about $1.5 billion per year was actually spent.

A: 80 years for the world and 60 years for the United States

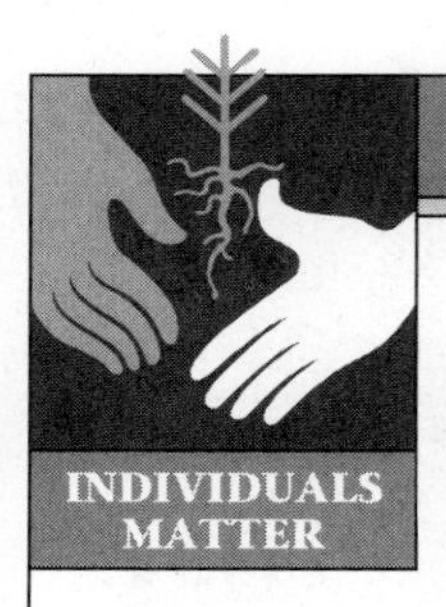

What *You* Can Do to Save Water (and Money)

- *For existing toilets, reduce the amount of water used per flush* by putting a tall plastic container weighted with a few stones into each tank, or by buying and inserting a toilet dam.
- *Install water-saving toilets that use no more than 6 liters (1.6 gallons) per flush.*
- *Flush toilets only when necessary.* Consider using the advice found on a bathroom wall in a drought-stricken area: "If it's yellow, let it mellow—if it's brown, flush it down."
- *Install water-saving showerheads and flow restrictors on all faucets.*
- *Check frequently for water leaks in toilets and pipes and repair them promptly.* A toilet must be leaking more than 940 liters (250 gallons) *per day* before you can hear the leak. To test for toilet leaks, add a water-soluble vegetable dye to the water in the tank, but don't flush. If you have a leak, some color will show up in the bowl's water within a few minutes.
- *Turn off sink faucets while brushing teeth, shaving, or washing.*
- *Wash only full loads of clothes*; if smaller loads must be washed, use the lowest possible water level setting.
- *When buying a new washer, choose one that uses the least amount of water and fills up to different levels for loads of different sizes.* Front-loading clothes models use less water and energy than comparable top-loading models.
- *Use automatic dishwashers for full loads only.* Also, use the short cycle and let dishes air-dry to save energy and money.
- *When washing many dishes by hand, don't let the faucet run.* Instead, use one filled dishpan or sink for washing and another for rinsing.
- *Keep one or more large bottles of water in the refrigerator rather than running water from the tap until it gets cold enough for drinking.*
- *Don't use a garbage disposal system*—a large user of water. Instead, consider composting your food wastes.
- *Wash a car from a bucket of soapy water and use the hose for rinsing only. Use a commercial car wash that recycles its water.*
- *Sweep walks and driveways instead of hosing them off.*
- *Reduce evaporation losses by watering lawns and gardens in the early morning or evening, rather than in the heat of midday or when it's windy.*
- *Use drip irrigation (Figure 7-9) and mulch for gardens and flower beds. Better yet, landscape with native plants adapted to local average annual precipitation so that watering is unnecessary or greatly reduced.*
- *To irrigate plants, install a system to capture rainwater or collect, filter, and reuse normally wasted gray water from bathtubs, showers, sinks, and the clothes washer.*

PRINCIPAL WATER POLLUTANTS The eight most common types of water pollutants are:

- *Disease-causing agents.* These include bacteria, viruses, protozoa, and parasitic worms that enter water from domestic sewage and animal wastes. In LDCs they are the biggest cause of sickness and death, prematurely killing an average of 13,700 people each day, half of them children under age 5.
- *Oxygen-demanding wastes.* These are organic wastes that can be decomposed by aerobic (oxygen-requiring) bacteria. Large populations of bacteria supported by these wastes can deplete water of dissolved oxygen, causing fish and other forms of oxygen-consuming aquatic life to die.
- *Water-soluble inorganic chemicals.* These consist of acids, salts, and compounds of toxic metals such as mercury and lead. High levels of these chemicals can make water unfit to drink, harm fish and other aquatic life, depress crop yields, and accelerate corrosion of equipment that uses the water.
- *Inorganic plant nutrients.* These are water-soluble nitrates and phosphates that can cause excessive growth of algae and other aquatic plants, which then die and decay, depleting water of dissolved oxygen and killing fish. Excessive levels of nitrates in drinking water can reduce the oxygen-carrying capacity of the blood of adults and can kill unborn children and infants, especially those under one year old.
- *Organic chemicals.* These include oil, gasoline, plastics, pesticides, cleaning solvents, detergents, and many other chemicals. They threaten human health and harm fish and other aquatic life.

Q: How long will the world's proven reserves of coal last at current consumption rates?

- *Sediment or suspended matter*. This is insoluble particles of soil and other solids that become suspended in water, mostly when soil is eroded from the land. By weight this is by far the biggest water pollutant. Sediment clouds water and reduces photosynthesis; it also disrupts aquatic food webs and carries pesticides, bacteria, and other harmful substances. Sediment that settles out destroys feeding and spawning grounds of fish, and it clogs and fills lakes, artificial reservoirs, stream channels, and harbors.
- *Radioactive isotopes* that are water-soluble or capable of being biologically amplified to higher concentrations as they pass through food chains and webs. Radiation from such isotopes can cause birth defects, cancer, and genetic damage.
- *Heat absorbed by water used to cool electric power plants*. The resulting rise in water temperature lowers dissolved oxygen content and makes aquatic organisms more vulnerable to disease, parasites, and toxic chemicals.

POINT AND NONPOINT SOURCES OF POLLUTION

Point sources discharge pollutants at specific locations through pipes, ditches, or sewers into bodies of surface water. Examples include factories, sewage treatment plants (which remove some but not all pollutants), active and abandoned underground mines, offshore oil wells, and oil tankers. Because point sources are at specific places (mostly in urban areas), they are fairly easy to identify, monitor, and regulate. In MDCs many industrial discharges are strictly controlled, whereas in LDCs such discharges are largely uncontrolled.

Nonpoint sources are sources that cannot be traced to any single discharge. They are usually large, poorly defined areas that pollute water by runoff, subsurface flow, or deposition from the atmosphere. Examples include runoff of chemicals into surface water and seepage into the ground from croplands, livestock feedlots, logged forests, streets, lawns, septic tanks, construction sites, parking lots, and roadways.

In the United States, nonpoint pollution from agriculture—mostly in the form of sediment, inorganic fertilizer, manure, salts dissolved in irrigation water, and pesticides—is responsible for an estimated 64% of the total mass of pollutants entering streams and 57% of those entering lakes. Little progress has been made in the control of nonpoint water pollution because of the difficulty and expense of identifying and controlling discharges from so many diffuse sources.

STREAM POLLUTION Flowing streams—including large ones called *rivers*—can recover rapidly from degradable, oxygen-demanding wastes and excess heat by a combination of dilution and bacterial decay (Figure 7-10). This recovery process works so long as streams are not overloaded with these pollutants and so long as their flow is not reduced by drought, damming, or diversion for agriculture and industry. Slowly degradable and nondegradable pollutants, however, are not eliminated by these natural dilution and degradation processes.

This breakdown of degradable wastes by bacteria depletes dissolved oxygen, which reduces or eliminates populations of organisms with high oxygen requirements until the stream is cleansed. The depth and width of the resulting *oxygen sag curve* (Figure 7-10)—and thus the time and distance a stream takes to recover—depend on the stream's volume, flow rate, temperature, and acidity, as well as the volume of incoming degradable wastes. Similar oxygen sag curves can be plotted when heated water from power plants is discharged into streams. The types of pollutants, flow rates, dilution capacity, and recovery time vary widely with different river basins; these factors also vary in the three major zones of a river as it flows from its headwaters, to its wider and deeper middle sections, and finally to an ocean or lake (Figure 7-4).

Requiring cities to withdraw their drinking water downstream rather than upstream (as is done now) would dramatically improve water quality as the stream flows toward the sea, for then each city would be forced to clean up its own waste outputs rather than pass them downstream. However, upstream users, who have the use of fairly clean water without high cleanup costs, fight this pollution prevention approach.

Water pollution control laws enacted in the 1970s have greatly increased the number and quality of wastewater treatment plants in the United States and in many other MDCs. Laws have also required industries to reduce or eliminate point-source discharges into surface waters. These efforts have enabled the United States to hold the line against increased pollution of most of its streams by disease-causing agents and oxygen-demanding wastes, an impressive accomplishment considering the rise in economic activity and population since the laws were passed.

However, we know relatively little about stream quality because water quality in 64% of the length of U.S. streams has not been measured. And many existing monitoring stations are not located in places that are suitable for assessing the presence or absence of pollutants from the surrounding drainage basins. Furthermore, even this limited monitoring does not measure concentrations of most toxic chemicals and ecological indicators of water quality.

Available data indicate that stream pollution from huge discharges of sewage and industrial wastes is a

A: 220 years (65 years if use increases by 2% a year)

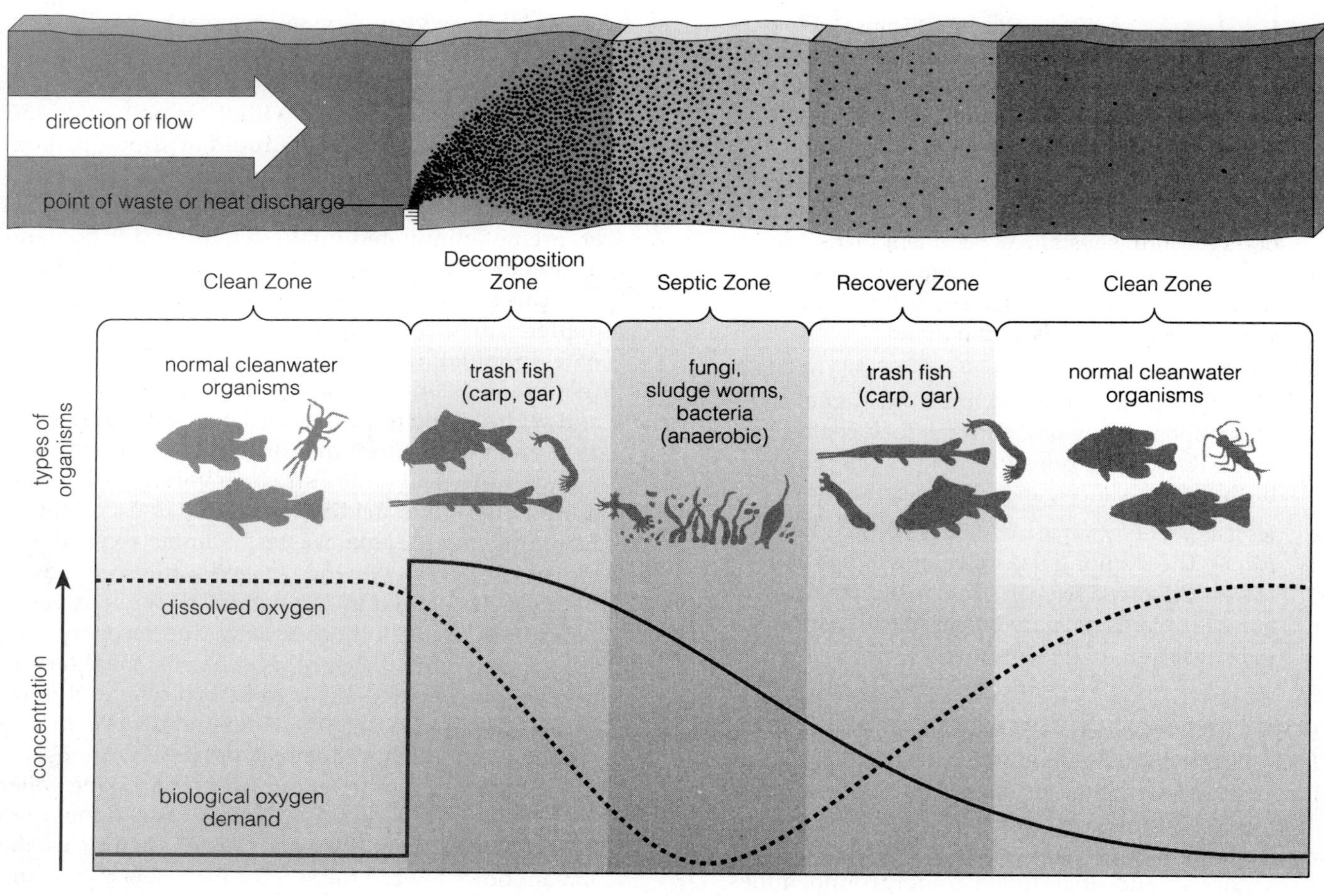

Figure 7-10 Dilution and decay of degradable, oxygen-demanding wastes and heat, showing the oxygen sag curve and the curve of oxygen demand. Depending on flow rates and the amount of pollutants, streams recover from oxygen-demanding wastes and heat if they are given enough time and are not overloaded.

serious and growing problem in most LDCs, where waste treatment is practically nonexistent. Numerous streams in the former Soviet Union and in eastern-European countries are severely polluted. Currently, more than two-thirds of India's water resources are polluted. Of the 78 streams monitored in China, 54 are seriously polluted. In Latin America and Africa, most streams passing through urban or industrial areas are severely polluted.

LAKE POLLUTION In lakes (and reservoirs), dilution is often less effective than in streams because these bodies of water frequently contain stratified layers that often undergo little vertical mixing (Figure 7-3). Stratification also reduces levels of dissolved oxygen, especially in the bottom layer. In addition, lakes and reservoirs have little flow, further reducing dilution and replenishment of dissolved oxygen. The flushing and changing of water in lakes can take from 1 to 100 years, compared with several days to several weeks for streams.

Thus lakes are more vulnerable than streams to contamination by plant nutrients, oil, pesticides, and toxic substances that can destroy bottom life and kill fish. Atmospheric fallout and runoff of acids is a serious problem in lakes vulnerable to acid deposition (Figure 6-10). Many toxic chemicals can also enter lakes from the atmosphere.

Lakes receive inputs of nutrients and silt from the surrounding land basin as a result of natural erosion and runoff. Near urban or agricultural areas, the input of nutrients to a lake can be greatly accelerated by human activities, a process known as *cultural eutrophication*. Such a change is caused mostly by nitrate- and phosphate-containing effluents from sewage treatment plants, runoff of fertilizers and animal wastes, and accelerated erosion of nutrient-rich topsoil.

 Q: What would happen if coal's harmful effects were included in its market price and government subsidies were removed?

During warm weather this nutrient overload produces dense growths of organisms such as algae, cyanobacteria, water hyacinths, and duckweed. Dissolved oxygen in both the surface layer of water near the shore and in the bottom layer is depleted when large masses of algae die, fall to the bottom, and are decomposed by aerobic bacteria. This oxygen depletion can kill fish and other oxygen-consuming aquatic animals. If excess nutrients continue to flow into a lake, the bottom water becomes foul and almost devoid of animals, as anaerobic bacteria take over and produce smelly decomposition products such as methane and highly toxic hydrogen sulfide.

About one-third of the 100,000 medium-to-large lakes and about 85% of the large lakes near major population centers in the United States suffer from some degree of cultural eutrophication. A quarter of China's lakes are classified as eutrophic.

The best solution to cultural eutrophication is to use both prevention methods to reduce the flow of nutrients into lakes and reservoirs and pollution cleanup methods to clean up lakes already suffering from excessive eutrophication. Major prevention methods include advanced waste treatment, bans or limits on phosphates in household detergents and other cleaning agents, and soil conservation and land-use control to reduce nutrient runoff. Major cleanup methods are dredging bottom sediments to remove excess nutrient buildup, removing excess weeds, controlling undesirable plant growth with herbicides and algicides, and pumping air through lakes and reservoirs to avoid oxygen depletion (an expensive and energy-intensive method).

OCEAN POLLUTION The oceans are the ultimate sink for much of the waste matter we produce. Oceans can dilute, disperse, and degrade large amounts of raw sewage, sewage sludge, oil, and some types of industrial waste, especially in deep-water areas. Marine life has also proved to be more resilient than some scientists had expected, leading some of them to suggest that it is generally safer to dump sewage sludge and most other hazardous wastes into the deep ocean than to bury them on land or burn them in incinerators.

Other scientists dispute this idea, pointing out that we know less about the deep ocean than we do about outer space. They add that dumping waste in the ocean would delay urgently needed pollution prevention activities and promote further degradation of this vital part of the earth's life-support system. Marine explorer Jacques Cousteau has warned that "the very survival of the human species depends upon the maintenance of an ocean clean and alive, spreading all around the world. The ocean is our planet's life belt."

Coastal areas—especially wetlands and estuaries, mangrove swamps, and coral reefs—bear the brunt of our enormous inputs of wastes into the ocean. This is not surprising, for half the world's population lives on the coast and another quarter lives within 80 kilometers (50 miles) of the sea.

In most coastal LDCs and in some coastal MDCs, untreated municipal sewage and industrial wastes are often dumped into the sea without treatment. In the United States, about 35% of all municipal sewage ends up virtually untreated in marine waters. Most U.S. harbors and bays are badly polluted from municipal sewage, industrial wastes, and oil.

Each year fully one-third of the area of U.S. coastal waters around the lower 48 states are closed to shellfish harvesters because of pollution and habitat disruption. In 1992 there were more than 2,600 beach closings in 22 coastal states, mostly because of bacterial contamination from inadequate and overloaded sewage treatment systems. Many more would be closed if their waters were tested regularly.

Dumping of industrial waste off U.S. coasts has stopped, although it still takes place in a number of other MDCs and some LDCs. However, barges and ships legally dump large quantities of *dredge spoils* (materials, often laden with toxic metals, scraped from the bottoms of harbors and rivers to maintain shipping channels) off the Atlantic, Pacific, and Gulf coasts at 110 sites.

In addition, many countries, including Great Britain, dump into the ocean large quantities of *sewage sludge*, a gooey mixture of toxic chemicals, infectious agents, and settled solids removed from wastewater at sewage treatment plants. This practice was banned in the United States as of 1992 by the Ocean Dumping Ban Act of 1988. Some elected officials and scientists oppose this ban, however, arguing that ocean disposal, especially in the deep ocean, is safer and cheaper than land dumping and incineration. Ships also dump large amounts of their garbage at sea because it is free; the alternative is to pay $500–$1,000 per ship for garbage disposal when they dock. An international ban on such dumping would be difficult to enforce.

Each year as many as 2 million seabirds and more than 100,000 marine mammals (including whales, seals, dolphins, sea lions, and sea turtles) die when they ingest or become entangled in plastic cups, bags, six-pack yokes, broken sections of fishing nets, ropes, and other debris dumped into the sea and discarded on beaches.

Since 1985, ocean dumping of radioactive waste in the open sea beyond the limits of national jurisdiction has been banned by an international agreement. However, Great Britain and Pakistan dispose of low-level radioactive wastes in coastal areas under their

A: It would be too expensive to use and would be phased out jurisdiction. And in 1992 it was learned that for decades the former Soviet Union had been dumping large quantities of high- and low-level radioactive wastes into the Arctic Ocean and tributaries that flow into this ocean.

Crude petroleum (oil as it comes out of the ground) and *refined petroleum* (fuel oil, gasoline, and other processed petroleum products) are accidentally or deliberately released into the environment from a number of sources. On March 24, 1989, the *Exxon Valdez*, a tanker more than three football fields long, went off course in a 16-kilometer-wide (7-mile) channel in Prince William Sound near Valdez, Alaska. It hit submerged rocks on a reef, creating the worst oil spill ever in U.S. waters. The rapidly spreading oil slick coated more than 1,600 kilometers (1,000 miles) of shoreline, almost the length of the shoreline between New Jersey and South Carolina. The oil killed between 300,000 and 645,000 birds (including 144 bald eagles), up to 5,500 sea otters, 30 seals, 23 whales, and unknown numbers of fish.

This multibillion-dollar accident might have been prevented if Exxon had spent $22.5 million to fit the tanker with a double hull. In the early 1970s, then-Interior Secretary Rogers Morton told Congress that all oil tankers using Alaskan waters would have double hulls, but under pressure from oil companies the requirement was later dropped. Today, virtually all merchant ships have double hulls—except oil tankers.

This spill highlighted the importance of pollution prevention. Even with the best technology and a fast response by well-trained people, scientists estimate that no more than 7–15% of the oil from a major spill can be recovered.

Tanker accidents and blowouts (oil escaping under high pressure from a borehole in the ocean floor) at offshore drilling rigs get most of the publicity, but more oil is released by normal operation of offshore wells, by washing tankers and releasing the oily water, and from pipeline and storage tank leaks.

Although natural oil seeps also release large amounts of oil into the ocean at some sites, most ocean oil pollution comes from activities on land, including pipeline and storage leaks and dumping of waste oil. A 1993 Friends of the Earth study estimated that each year U.S. oil companies unnecessarily spill, leak, or waste oil equal to that held by 1,000 *Exxon Valdez* tankers, or more oil than Australia uses. Almost half (some experts estimate 90%) of the oil reaching the oceans is waste oil dumped onto the land or into sewers by cities, individuals, and industries. Each year oil equal to 20 times the amount spilled by the *Exxon Valdez* is improperly disposed of by U.S. citizens changing their own motor oil.

Research shows that most forms of marine life recover from exposure to large amounts of crude oil within three years. However, recovery from exposure to refined oil, especially in estuaries, may take 10 years or longer. The effects of spills in cold waters (such as Alaska's Prince William Sound and Antarctic waters) and in shallow enclosed gulfs and bays (such as the Persian Gulf) generally last longer.

Oil slicks that wash onto beaches can have a serious economic impact on coastal residents, who lose income from fishing and tourist activities. Oil-polluted beaches washed by strong waves or currents are cleaned up after about a year, but beaches in sheltered areas remain contaminated for several years. Estuaries and salt marshes suffer the most damage and cannot effectively be cleaned up.

GROUNDWATER POLLUTION While highly visible oil spills get lots of media attention, a much greater threat to human health is the out-of-sight pollution of groundwater, which is a prime source of water for drinking and irrigation. This vital form of Earth capital is easy to deplete and pollute because it is renewed so slowly. Laws protecting groundwater are weak in the United States and nonexistent in most countries.

When groundwater becomes contaminated, it does not cleanse itself of degradable wastes, as surface water can if it is not overloaded. Because groundwater flows are slow and not turbulent, contaminants are not effectively diluted and dispersed. Also, groundwater has much smaller populations of decomposing bacteria than do surface water systems, and its cold temperature slows down decomposition reactions. Thus it can take hundreds to thousands of years for contaminated groundwater to cleanse itself of degradable wastes—and nondegradable wastes are there permanently.

Results of limited testing of groundwater in the United States are alarming. In a 1982 survey, the EPA found that 45% of the large public water systems served by groundwater were contaminated with synthetic organic chemicals that posed potential health threats. Another EPA survey in 1984 found that two-thirds of the rural household wells tested violated at least one federal health standard for drinking water, usually pesticides or nitrates from fertilizers (which cause a life-threatening blood disorder in infants during their first year). The EPA has documented groundwater contamination by 74 pesticides in 38 states.

Crude estimates indicate that although only 2% by volume of all U.S. groundwater is contaminated, up to 25% of usable groundwater is contaminated, and in some areas as much as 75% is contaminated. In New Jersey, for example, every major aquifer is contaminated. In California, pesticides contaminate the drinking

Q: Has a scientifically and politically acceptable method for the long-term disposal of nuclear waste been developed?

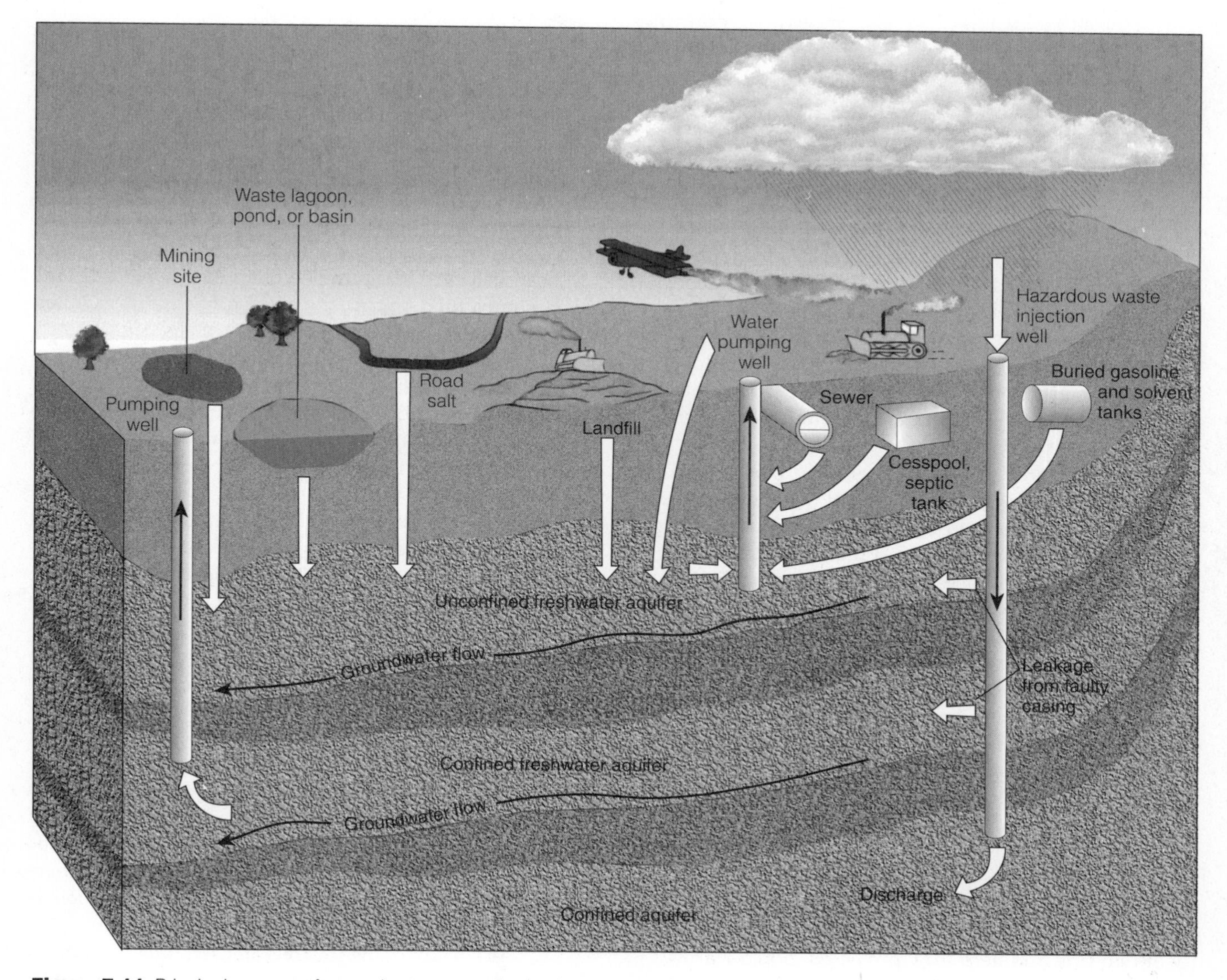

Figure 7-11 Principal sources of groundwater contamination in the United States.

water of more than 1 million people. In Florida, where 92% of the residents rely on groundwater for drinking, over 1,000 wells have been closed.

Groundwater can be contaminated from a number of sources, including underground storage tanks, landfills, abandoned hazardous-waste dumps, deep wells used to dispose of liquid hazardous wastes, and industrial-waste storage lagoons located above or near aquifers (Figure 7-11). An EPA survey found that one-third of 26,000 industrial-waste ponds and lagoons have no liners to prevent toxic liquid wastes from seeping into aquifers. One-third of those sites are within 1.6 kilometers (1 mile) of a drinking-water well.

The EPA estimates that at least 1 million underground tanks are leaking their contents into groundwater. A slow gasoline leak of just 4 liters (1 gallon) per day can seriously contaminate the water supply for 50,000 people. Such slow leaks usually remain undetected until someone discovers that a well is contaminated.

Determining the extent of a leak can cost $25,000–-$250,000. Cleanup costs range from $10,000 for a small spill to $250,000 or more if the chemical reaches an aquifer. Legal fees and damages to injured parties can run into the millions. Stricter regulations should reduce leaks from new tanks but would do little about the millions of older tanks that are "toxic time bombs." Some analysts call for above-ground storage of hazardous liquids so that leaks can be more easily detected and rectified.

A: No

7-5 Solutions: Preventing and Controlling Water Pollution

NONPOINT-SOURCE POLLUTION The leading nonpoint source of water pollution is agriculture. Farmers can sharply reduce fertilizer runoff into surface waters and leaching into aquifers by using moderate amounts of fertilizer, and by using none at all on steeply sloped land. They can use slow-release fertilizers and alternate their plantings between row crops and soybeans or other nitrogen-fixing plants to reduce the need for fertilizer. Farmers can also be required to plant buffer zones of permanent vegetation between cultivated fields and nearby surface water.

Farmers can also reduce pesticide runoff and leaching by applying pesticides only when needed. They can reduce the need for pesticides by using biological control or integrated pest management (Solutions, p. 51). Nonfarm uses of inorganic fertilizers and pesticides—on golf courses, yards, and public lands, for example—could also be sharply reduced.

Livestock growers can control runoff and infiltration of manure from feedlots and barnyards by managing animal density, by planting buffers, and by not locating feedlots on land that slopes toward nearby surface water. Diverting the runoff into detention basins would allow this nutrient-rich water to be pumped and applied as fertilizer to cropland or forestland.

Critical watersheds should also be reforested. Besides reducing water pollution from sediments, reforestation would reduce soil erosion and the severity of flooding and help slow projected global warming (Section 6-2) and loss of the earth's vital biodiversity (Chapter 4).

POINT-SOURCE POLLUTION In many LDCs and in some MDCs, sewage and waterborne industrial wastes are discharged without treatment into the nearest waterway or into wastewater lagoons. In Latin America, less than 2% of urban sewage is treated. Only 15% of the urban wastewater in China receives treatment. Treatment facilities in India cover less than a third of the urban population.

In MDCs, most wastes from point sources are purified to varying degrees. Between 1972 and 1992, U.S. taxpayers and the private sector have spent more than $541 billion on water pollution control—nearly all of it on end-of-pipe controls on municipal and industrial discharges from point sources mandated by water pollution control laws.

In rural and suburban areas with suitable soils, sewage from each house is usually discharged into a *septic tank*. Such a system traps greases and large solids in an underground concrete settling tank and uses an array of buried perforated pipes to discharge the remaining wastes over a large drainage field. As these wastes percolate downward, the soil filters out some potential pollutants, and soil bacteria decompose biodegradable materials. To be effective, septic tank systems must be properly installed in soils with adequate drainage, not placed too close together or too near well sites, and pumped out when the settling tank becomes full. About 25% of all homes in the United States are served by septic tanks.

In urban areas, most waterborne wastes from homes, businesses, factories, and storm runoff flow through a network of sewer pipes to wastewater treatment plants. When sewage reaches a treatment plant, it can undergo up to three levels of purification, depending on the type of plant and the degree of purity desired. *Primary sewage treatment* is a mechanical process that uses screens to filter out debris such as sticks, stones, and rags. Then suspended solids settle out as sludge in a settling tank (Figure 7-12).

Secondary sewage treatment is a biological process in which aerobic bacteria are used to remove up to 90% of biodegradable, oxygen-demanding organic wastes (Figure 7-12). After secondary treatment, however, wastewater still contains about 3–5% by weight of the oxygen-demanding wastes, 3% of the suspended solids, 50% of the nitrogen (mostly as nitrates), 70% of the phosphorus (mostly as phosphates), and 30% of most toxic metal compounds and synthetic organic chemicals. Virtually none of any long-lived radioactive isotopes or persistent organic substances such as pesticides is removed.

As a result of the Clean Water Act, most U.S. cities have secondary sewage treatment plants. In 1989, however, the EPA found that more than 66% of sewage treatment plants have water-quality or public-health problems. Also, 500 cities have failed to meet federal standards for sewage treatment plants, and 34 East Coast cities simply screen out large floating objects from their sewage before discharging it into coastal waters.

Advanced sewage treatment is a series of specialized chemical and physical processes that remove specific pollutants left in the water after primary and secondary treatment. Types of advanced treatment vary depending on the specific contaminants to be removed. Advanced treatment is rarely used because such plants typically cost twice as much to build and four times as much to operate as secondary plants. However, despite the cost, advanced treatment is used for more than a third of the population in Finland, the former West Germany, Switzerland, and Sweden, and to a lesser degree in Denmark and Norway.

Q: Does using nuclear power add carbon dioxide to the atmosphere?

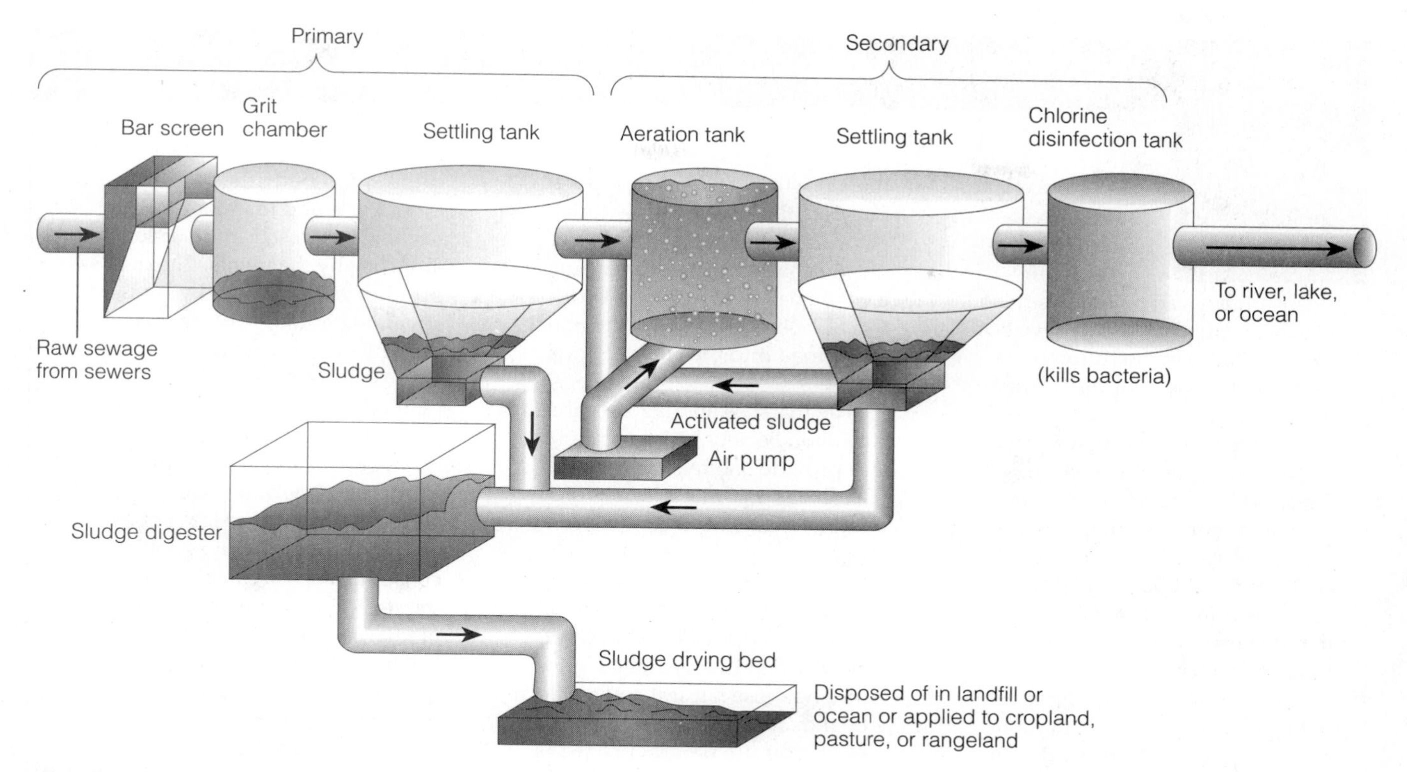

Figure 7-12 Primary and secondary sewage treatment.

Before water is discharged after primary, secondary, or advanced treatment, it is bleached to remove water coloration and disinfected to kill disease-carrying bacteria and some, but not all, viruses. The usual method for doing this is *chlorination*. However, chlorine reacts with organic materials in water to form small amounts of chlorinated hydrocarbons, some of which cause cancers in test animals. Disinfectants such as ozone and ultraviolet light are being used in some places, but they cost more than chlorination.

Primary and secondary sewage treatment produces a toxic gooey sludge that must be disposed of or recycled to the land as fertilizer. About 54% by weight of all municipal sludge produced in the United States is applied to farmland, forests, highway medians, and degraded land as fertilizer, and 9% is composted. The rest is dumped in conventional landfills (where it can contaminate groundwater) or incinerated (which can pollute the air with traces of toxic chemicals, and the resulting toxic ash is usually buried in a landfill that will eventually leak).

Before it is applied to land, sewage sludge can be heated to kill harmful bacteria, as is done in Switzerland and parts of Germany. It can also be treated to remove toxic metals and organic chemicals before application, but that can be expensive. The best and cheapest solution is to prevent these toxics from reaching sewage treatment plants. However, untreated sludge can be applied to land not used for crops or livestock. Examples include forests, surface-mined land, golf courses, lawns, cemeteries, and highway medians.

It is encouraging that some communities and individuals are seeking better ways to purify contaminated water by working with nature (Solutions, p. 148).

PROTECTING DRINKING WATER Treatment of water for drinking by urban residents is much like wastewater treatment. Areas that depend on surface water usually store it in a reservoir for several days to improve clarity and taste by allowing the dissolved oxygen content to increase and suspended matter to settle out. The water is then pumped to a purification plant, where it is treated to meet government drinking water standards. Usually, it is run through sand filters, then through activated charcoal, and then it is disinfected. In areas with very pure sources of groundwater, little, if any, treatment is necessary.

Only about 54 countries, most of them in North America and Europe, have safe drinking water standards. The Safe Drinking Water Act of 1974 requires the EPA to establish national drinking water standards, called *maximum contaminant levels*, for any pollutants that may have adverse effects on human health. This act has helped improve drinking water in much of the United States, but much more still needs to be done. More

A: Yes, but it produces only one-sixth as much per unit of electricity as a coal-burning plant

Working with Nature to Purify Sewage

SOLUTIONS

Natural wetlands have a great—but not unlimited—capacity to cleanse. They can be used to treat urban sewage, but many have been overwhelmed by pollution or destroyed by development. An exciting low-tech, low-cost alternative to expensive waste treatment plants is to create an artificial wetland, as the residents of Arcata, California, did on land that was once a dump between the town and adjacent Humboldt Bay. The project was completed in 1974 for $3 million less than the estimated cost of a conventional treatment plant.

Here's how it works: First, sewage is held in sedimentation tanks, where the solids settle out. This resulting sludge is removed and processed for use as fertilizer. The liquid is pumped into oxidation ponds, where the wastes are broken down by bacteria. After a month or so, the water is released into the artificial marshes, where it is further filtered and cleansed by plants and bacteria. Although the water is clean enough to discharge directly into the bay, state law requires that it first be chlorinated. So the town chlorinates the water—and then dechlorinates it—before sending it into the bay, where oyster beds thrive. Some water from the marshes is piped into the city's salmon hatchery.

The marshes and lagoons are an Audubon Society bird sanctuary and provide habitats for thousands of seabirds and marine animals. The treatment center is a city park and attracts many tourists. The town even celebrates its natural sewage-treatment system with an annual "Flush with Pride" festival. Over 150 cities and towns in the United States now use natural and artificial wetlands for treating sewage.

Can you use natural processes for treating wastewater if there isn't a wetland available, or enough land on which to build one? According to ecologist John Todd, you can: Set up a greenhouse lagoon and use sunshine the way nature does. The process begins when sewage flows into a greenhouse containing rows of large aquarium tanks covered with plants such as water hyacinths, cattails, and bulrushes. In these tanks algae and microorganisms decompose wastes into nutrients absorbed by the plants. The decomposition is speeded up by sunlight streaming into the greenhouse, and toxic metals are absorbed into the tissues of trees that will be transplanted outside. Then the water passes through an artificial marsh of sand, gravel, and bulrush plants that filters out algae and organic waste. Next the water flows into aquarium tanks, where snails and zooplankton consume microorganisms and are themselves consumed by crayfish, tilapia, and other fish that can be eaten or sold as bait. After 10 days the now-clear water flows into a second artificial marsh for final filtering and cleansing. When working properly, such solar-aquatic treatment systems have produced water fit for drinking.

These natural alternatives to building expensive treatment plants may not solve the waste problems of large cities, but they can help, and they are an attractive alternative for small towns, the edges of urban areas, and rural areas.

money is spent on military bands each year than on the EPA's enforcement of the Safe Drinking Water Act.

Privately owned wells in suburban and rural areas are not required to meet federal drinking water standards. The biggest reasons are the cost of testing each well regularly (at least $1,000) and ideological opposition to mandatory testing and compliance by some homeowners.

In 1993 a study by the Natural Resources Defense Council found that 43% of all U.S. municipal drinking water systems violated one or more drinking water standards during 1991. These violations affected more than 120 million people and caused an estimated 900,000 illnesses and 900 deaths. In most cases, people were not notified when their drinking water was contaminated.

Contaminated wells and concern about possible contamination of public drinking water supplies have created a boom in the number of U.S. citizens drinking bottled water at costs about 1,500 times more than that of tap water, or of those adding water purification devices to their home systems. This has created enormous profits for both legitimate companies and con artists in these businesses. Many bottled-water drinkers are getting ripped off. More than one-third of the bottled water comes from the same sources used to supply tap water, which is regulated much more strictly than bottled water.

To be safe, consumers should purchase bottled water only from companies that have their water frequently tested and certified, ideally by EPA-certified laboratories. Before buying bottled water the con-

Q: Can switching to increased use of nuclear power in the United States save much oil?

sumer should determine whether the bottler belongs to the International Bottled Water Association (IBWA) and adheres to its testing requirements. The IBWA requires its members to test for 181 contaminants and sends an inspector from the National Sanitation Foundation, a private lab, to bottling plants annually to check all pertinent records and make sure the plant is run cleanly.

PROTECTING COASTAL WATERS The most important suggestions for preventing excessive pollution of coastal waters and for cleaning them up include the following:

Prevention

- *Greatly reducing the discharge of toxic pollutants into coastal waters from both industrial facilities and municipal sewage treatment plants.*
- *Greatly reducing all discharges of raw sewage from sewer-line overflows by requiring separate storm and sewer lines in cities.*
- *Banning all ocean dumping of sewage sludge and hazardous dredged materials.*
- *Enacting and enforcing laws and land-use practices that sharply reduce runoff from nonpoint sources in coastal areas.*
- *Protecting sensitive marine areas from all development by designating them as ocean sanctuaries.*
- *Regulating coastal development to minimize its environmental impact, and eliminating subsidies and tax incentives that encourage harmful coastal development.*
- *Prohibiting oil drilling in ecologically sensitive offshore and nearshore areas.*
- *Collecting used oils and greases from service stations and other sources, and reprocessing them for reuse.* Currently, more than 90% of the used oil collected for "recycling" is burned as fuel, releasing lead, chromium, arsenic, and other pollutants into the air.
- *Requiring all existing oil tankers to have double hulls, double bottoms, or other oil-spill prevention measures by 1998.*
- *Greatly increasing the financial liability of oil companies for cleaning up oil spills, thus encouraging pollution prevention.*
- *Routing oil tankers as far as possible from sensitive coastal areas.*
- *Having Coast Guard or other vessels guide tankers out of all harbors and enclosed sounds and bays.*
- *Banning the rinsing of empty oil tanker holds and the dumping of oily ballast water into the sea.*
- *Banning discharge of garbage from vessels into the sea and levying large fines on violators.*
- *Adopting a nationwide tracking program to ensure that medical waste is safely disposed of.*

Cleanup

- *Greatly improving oil-spill cleanup capabilities.* However, according to a 1990 report by the Office of Technology Assessment, there is little chance that large spills can be effectively contained or cleaned up.
- *Upgrading all coastal sewage treatment plants to at least secondary treatment, or developing alternative methods for sewage treatment* (Solutions, left).

PROTECTING GROUNDWATER Pumping polluted groundwater to the surface, cleaning it up, and returning it to the aquifer is usually prohibitively expensive—$5–10 million or more for a single aquifer. Recent attempts to pump and treat contaminated aquifers show that it may take decades, even hundreds of years, of pumping before all of the contamination is forced to the surface. Researchers, however, are working to find natural bacteria or to develop genetically engineered bacteria that can degrade specific pollutants found in groundwater.

Clearly, preventing contamination is the most effective way to protect groundwater resources. Water pollution experts suggest that this could be accomplished by the following means:

- *Banning virtually all disposal of hazardous wastes in sanitary landfills and deep injection wells*
- *Monitoring aquifers near existing sanitary and hazardous-waste landfills, underground tanks, and other potential sources of groundwater contamination*
- *Controlling application of pesticides and fertilizers by farmers and homeowners more strictly*
- *Establishing pollution standards for groundwater*
- *Emphasizing above-ground storage of hazardous liquids—an in-sight-in-mind approach that allows rapid detection and collection of leaks*

Each of us has a role to play in reducing water waste (Individuals Matter, p. 140) and pollution (Individuals Matter, p. 150).

It is not until the well runs dry, that we know the worth of water.

BENJAMIN FRANKLIN

A: No. In 1992, burning oil supplied only 3% of the country's electricity.

What *You* Can Do to Reduce Water Pollution

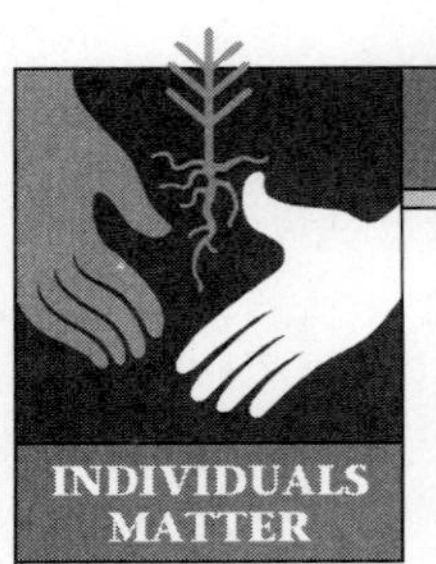

- *Use manure or compost instead of commercial inorganic fertilizers on garden and yard plants.*
- *Use biological methods or integrated pest management* (Solutions, p. 51) *instead of commercial pesticides to control garden, yard, and household pests.*
- *Use low-phosphate, phosphate-free, or biodegradable dishwashing liquid, laundry detergent, and shampoo.*
- *Don't use water fresheners in toilets.*
- *Use less-harmful substances, such as baking soda, vinegar, and borax, instead of commercial chemicals for most household cleaners.*
- *Don't pour pesticides, paints, solvents, oil, or other products containing harmful chemicals down the drain or on the ground.* Contact your local health department about disposal.
- *If you get water from a private well or suspect that municipal water is contaminated, have it tested by an EPA-certified laboratory for lead, nitrates, trihalomethanes, radon, volatile organic compounds, and common pesticides.*
- *If you have a septic tank, have it cleaned out every three to five years by a reputable contractor so that it won't contribute to groundwater pollution.*
- *Get to know your local bodies of water and form community watchdog groups to help monitor, protect, and restore them.*
- *Support efforts to clean up riverfronts and harbors.*

CRITICAL THINKING

1. How do human activities contribute to flooding? How can these effects be reduced?
2. Should prices of water for all uses in the United States be raised sharply to encourage water conservation? Explain. What effects might this have on the economy, on you, on the poor, and on the environment?
3. Why is dilution not always the solution to water pollution? Give examples and conditions for which this solution is, and is not, applicable.
4. Should all dumping of wastes in the ocean be banned? Explain. If so, where would you put the wastes instead? What exceptions would you permit, and why?
5. Because the deep oceans are vast and are located far away from human habitats, why not use them as the depository for our radioactive and other hazardous wastes? Give your reasons for agreeing or disagreeing with this proposal.

Index

Page numbers appearing in **boldface** indicate where definitions of key terms can be found in the text. Page numbers in *italics* indicate illustrations, tables, and figures.

S

T

U

W